King of the Crocodylians

LIFE OF THE PAST
James O. Farlow, editor

King of the Crocodylians

The Paleobiology of *Deinosuchus*

David R. Schwimmer

INDIANA
University Press
Bloomington and Indianapolis

This book is a publication of

Indiana University Press
601 North Morton Street
Bloomington, IN 47404-3797 USA

http://iupress.indiana.edu

Telephone orders 800-842-6796
Fax orders 812-855-7931
Orders by e-mail iuporder@indiana.edu

CIP and print line to come

Library of Congress Cataloging-in-Publication Data

Schwimmer, David R.
    King of the crocodylians : the paleobiology of deinosuchus / David R. Schwimmer.
        p. cm. — (Life of the past)
    Includes bibliographical references and index.
    ISBN 0-253-34087-X (cloth : alk. paper)
    1. Deinosuchus.   I. Title. II. Series.
QE862.C8 S38 2002
567.9'8—dc21   2001005660

1   2   3   4   5   07   06   05   04   03   02

CONTENTS

For many years, dating back to my early childhood, a huge, unbeliev-ably scary, fossil "crocodile" skull occupied a glass-walled corner of the fourth floor of the American Museum of Natural History in New York. Millions of tourists, local New Yorkers (like me), paleontologists (as I wanted to be), teachers, and students have stared at the skull and felt the dread that it transmitted through the glass. On the back wall of the display was a mural depicting the habitat and prey of the fearful giant, painted in classic 1950s dinosaur-tableau style. It featured an anxious horned dinosaur in the foreground and several dithering duck-billed dinosaurs in the distance, all clearly cowering from the crocodylian giant represented by the awesome skull.

The skull was enormous in all dimensions. It was broad at the rear and high; resting on the display floor, the height of the skull was about 1.2 m, which was roughly my height when I first saw it. The skull had the jaws gaping open in a position of maximum menace, showing teeth about 10 cm long, all sharply pointed. This was the image of *Deino-suchus* at that time (identified in the 1950s as "*Phobosuchus*"), the largest land carnivore in North America of the late Age of Dinosaurs—and certainly the largest crocodylian known from this continent. Our image of *Deinosuchus* has changed somewhat over the past 45 years, but it was still among the largest and most formidable land carnivores of all time.

This book is an ecological study of *Deinosuchus*, which is an idea itself that may require some explanation. Despite the familiar modern usage of the term ("We must save the ecology!"), "ecology" literally means in Greek "the study of the house" (*eko*, house, and *logos*, study). In everyday speech, the word is often used as a noun: we might refer to the "study of an ecology" (or some similar phrase). If we study an ancient creature, it is properly termed "paleoecology" (*paleo*, ancient). There are many ways to approach such studies. If one examines a whole ecosystem and considers the roles of all the players (for example, we can study the coral reef ecosystem and consider the contributions of coral, fish, and sponges), this would be a "synecology" (*syn*, together). And if such a study were dealing with an ancient reef community, it would be a "paleosynecology." But if we study the life, times, and habitat of a single organism, it is called an "autecology" (*auto*, alone), and when it deals with a creature of the past, it becomes a "paleoautecology."

With these terms clarified, then, I can state precisely that this book on *Deinosuchus* is a paleoautecology of one of the most intriguing (and strangely overlooked) giant animals of the late Mesozoic Era. These huge animals were the dominant predators along the southern marine coasts of North America for roughly 10 million years during the later Cretaceous Period. As I will explain in detail to follow, in the coastal regions where *Deinosuchus* reigned, even the carnivorous theropod dinosaurs were its prey.

The reader will notice that I use the term "crocodylian" to indicate the general classification of *Deinosuchus*. Why not use "crocodile" or at least a more familiar general term? For arcane taxonomic reasons (discussed in Chapter 7), a "crocodile" in modern usage refers only to the many species of the modern and near-modern genus *Crocodylus* (including such well-known species as the Nile crocodile, *Crocodylus niloticus,* and the Indo-Pacific crocodile, *C. porosus*). The new information that has been gathered about *Deinosuchus* and that is presented in many parts of this book shows that it is not close to the ancestry of modern crocodiles, and so I can't comfortably use that familiar term as a general reference. Despite some superficial differences in the skull, *Deinosuchus* turns out to be close to the ancestry of the alligators (discussed below), but it is still too far from the more evolved alligators to allow me to comfortably use that term either. If a creature belongs to the generalized group of crocodilelike animals but it's really a very basal alligator relative, what are we to call it? The older generalized term "crocodilian" might work, but it is derived from a misspelling, because as stated, the genus of true crocodiles is *Crocodylus*. (Thus, a *Crocodylus* species is a legitimate "crocodile" but it should really be spelled "crocodyle.") To resolve the minor terminology dilemma, modern specialists who deal with these animals have adopted the term "crocodylians" in reference to the group that includes modern and fossil true crocodiles, alligators, gharials, and their closer relatives.

In fact, I will also show that modern studies reveal some surprising relationships among the long-snouted, aquatic animals we envision with terms such as "crocodylian." It turns out that the true giants among them, which reached up to 12 m and well over 8 tons, tended to occur among forms closer to the alligators rather than to the crocodiles. And in that fashion, *Deinosuchus* fits well among the earliest alligatoroids, which includes the higher crocodylians—those closer to the alligators than to the crocodiles or gharials. This identification as an alligatoroid may surprise people generally familiar with the concept of *Deinosuchus,* especially because its tapered, constricted snout in the famous skull in the American Museum of Natural History somehow *looks* like a crocodile's. But that snout constriction is a basal character of many fossil crocodylians, which just happened to become lost in modern alligators: only in the modern world does it indicate a difference between crocodiles and alligators. Therefore, *Deinosuchus* cannot be called a "crocodile," or a "crocodilian," or an "alligator," and for the sake of clarity, I must burden the reader with the unfamiliar term "crocodylian" as the generalized adjective and noun for the larger group including *Deinosuchus*. Forgive the unfamiliar term: these creatures are worth the effort.

Acknowledgments

As with any work of this nature, many people have provided ideas, images, data, or valuable conversations that make the results possible. Pity the writer who must list the contributors in some sort of meaningful order when there are so many to acknowledge. Doing so in approximate chronological order, I begin with thanks to the National Geographic Society, who provided funds that began my study of *Deinosuchus* and led to this book. Their funding allowed me to travel to the many localities where *Deinosuchus* fossils appear and to the institutions where they are housed. Next came Bob Sloan, senior sponsoring editor at Indiana University Press (IUP), and Jim Farlow, the series editor for the Life of the Past series at IUP, who directly began the process leading to this book. Both Bob and Jim have proven to be the most unmeddlesome of editors, providing help when asked and just that.

Among the colleagues who shared ideas and materials with me, Wann Langston Jr. of the University of Texas, Austin, is clearly foremost. Professor Langston is the leading exponent of the study of *Deinosuchus,* and he has unselfishly shared his new and emerging information with me. Among his students was Chris Brochu, who himself has become one of the most significant researchers on fossil crocodylians (as the reference listing shows), and I appreciate Chris's insights, which he freely shared with me. Gregory Erickson is a colleague who has specialized in the biomechanics of large fossil animals, and his work (including several collaborations with Chris Brochu) and conversations with me formed the underpinnings for some of the critical ideas developed here. I also greatly appreciate the conversations and specimens provided by Professor Dale Russell of North Carolina State University in Raleigh, James Lamb at the same institution, and David Parris and Barbara Grandstaff of the New Jersey State Museum, Trenton. Additional specimens were offered for study by David Dockery of the Mississippi State Geological Survey, Jackson, Ted Daeschler of the Academy of Natural Sciences, Philadelphia, and Charlotte Holton of the American Museum of Natural History, New York.

My work over the past two decades has been possible only with the assistance and cooperation of amateur collectors, and two stand out in particular. G. Dent Williams of Seale, Alabama, collected many of the specimens that led to the ideas in this book and supplied field assistance and insights into the detailed skull anatomy of our southeastern *Deino-suchus* specimens. Ken Barnes, of Terlingua, Texas, kindly allowed me to examine and photograph his crocodylian-bitten dinosaur bones from the Big Bend region, a subject of great interest that I cover in Chapter 8.

The photographs and graphics in this book are partly my work, but mostly the work of others. D. W. Miller painted the cover art, carefully following my specifications for the anatomy of *Deinosuchus* and using contemporary birds and fish and other ancient life for backdrops. Ron Hirzel created most of the pen-and-ink drawings throughout the book, also carefully reconstructing the anatomy and scenes I requested. All their work included here was created specifically for this book. My colleague Bill Frazier of Columbus State University (CSU), Columbus, Georgia, also provided several figures used here. Several photographs, as well as extensive proofreading, were the work of Tracy Hall of CSU, and most of the graphics reproduction work, as well as all the computer composites, performed by Jon Haney of the CSU Instructional Technologies Laboratory. Photographs that are not otherwise labeled are mine. I applaud and thank all of my editors, mentors, colleagues, friends, students, and artists for their talents and their help.

King of the Crocodylians

1. The Life and Times of a Giant Crocodylian

An Encounter at the Southern Shore

It is late autumn, many years in the past in the area of Lowndes County, close to Montgomery, central Alabama. The weather is quite warm, with daytime air temperatures around 35°C (95°F), in part because at this time, this site lies on the Gulf of Mexico seashore, receiving climate-moderating winds coming from the relatively warm ocean. It is also warm because the prevailing global climate is warm, with less seasonal variation than would come about later in geological time during and following the series of episodes commonly called the Ice Age.

The shoreside plant life within sight is generally familiar (Fig. 1.1): pine trees make up dense groves on upshore sandy patches, reaching down in places to the upper tide zone of the beach. Dune tracts are scattered among the pine groves, held together with tall *Phragmites* reeds (which are among the first grasses of the fossil record). Farther inland are dense groves of *Podocarpus,* a primitive gymnospermous tree: these later became extinct in North America, but they survive today in South America. Wetter areas above the shoreface contain abundant, low-growing lycopods (scouring rushes), ferns, and herbaceous cycads. A few very tall trees may be seen scattered throughout the shore groves: these are sequoias, which later in geological time become restricted to small areas of North America and Asia. Where muddy deposits reach close or to the wave zone, the sandy beach gives way to dense mangroves, which form a barrier to the sea with their buttress roots covered by oysters (as well as some less familiar bivalve mollusks). Upshore from the beach and shore groves are vast salt marshes. These marshes run parallel to the shoreline, extending many tens of kilometers landward because of the extremely low gradient (slope) of

Figure 1.1. Salt marsh and beach dune setting in the center of Late Cretaceous Alabama. Drawing by Ron Hirzel.

the land surface, where both runoff from the land and frequent storm surges combine to keep water tables high. As a result, vast areas of the southeastern coast of North America are covered by shallow, brackish standing water.

The salt marshes, beach, and nearshore waters support abundant life. In the distance, wading through the salt marsh and feeding on the rich vegetation in and around it is a small herd of *Lophorhothon,* herbivorous iguanodontid dinosaurs who trumpet to each other through Roman-nosed snouts. This herd contains mostly adults, with body lengths to 9 m and weights averaging 2.5 tons. Out in the open ocean, a *Platecarpus* mosasaur pursues a school of large tarponlike *Xiphactinus* fish, which, at up to 4 m length themselves, are usually the predators. Under the waves, on the bottom and midwaters of the shallow continental shelf, swim and forage hosts of rays, sawfish, galeomorph sharks (that is, typical fast-swimming forms), and smaller bony fish. A peculiar, large *Scapanorhynchus* shark, with its odd, elongate rostrum and a nearly single-lobed tail, patrols the schools of small fish, selecting those it can catch as prey. It is an ambush predator and can only achieve short bursts of speed for prey capture, but the waters are full of life and it finds ample food.

Elsewhere along the shore and in the air, still more life is evident. A few medium-sized pterosaurs, with 3-m wingspans, of a genus that paleontologists in modern times still have not yet identified, soar above the sea, picking off schooling *Enchodus* fish that venture close to the surface. Many small and a few larger birds occupy the marshes and nearshore area, feeding in a similar fashion on yet smaller fish, crabs, and insects. The most common shorebirds are robin-sized *Ichthyornis*. Much larger, much more rare *Baptornis* paddle far offshore, diving for *Enchodus* in the manner of latter-day cormorants. These *Ichthyornis* and *Baptornis* birds appear quite ordinary externally, except that each has a beak full of teeth, and the *Baptornis* has no visible trace of wings.

In the near distance, a young, male theropod (carnivorous) dinosaur, who is really another form of bird, strolls the shoreface looking for smaller dinosaurs, snakes, lizards, or mammals, or for something to scavenge (Fig. 1.2). It belongs to an as-yet-unnamed genus in a primitive branch of the tyrannosauroids, and it will reach an adult weight of about 2 tons and a length of 8 m. But for now, this teenage theropod is off on its own, having left its home nest, located far to the north, well beyond the area of the salt marshes. This theropod weighs perhaps 600 kg and is 6 m in length—about the size of a very large modern horse, but much longer and leaner, all claws and teeth. It will rapidly grow closer to its adult size if it finds sufficient food and survives. The *Lophorhothon* herd in the distance has animals too large and well protected for a kill by the young carnivore, and so it wanders fruitlessly along the dune area and then heads up shore toward the salt marsh. It sees a smaller *Lophorhothon* alone in the distance, perhaps itself having wandered from the herd, wading through the marsh weeds, and the predator begins to circle around the edge of the salt marsh, partly hidden in the *Podocarpus* groves. The theropod emerges from the trees and wades out into marsh to cut off the prey dinosaur, verging perhaps 50 m out from the hard ground surface. The mucky water reaches more than a meter in depth, and the soft bottom sediment engulfs even more of the theropod's legs. But his legs are 2.5 m long, and he can swim if necessary, so he continues across the marsh.

Suddenly, an enormous mass explodes from the bottom of the salt marsh, covering the 10-m distance to the theropod in three seconds. Jaws nearly a meter long open wide and close on the lower back of the theropod, which lies close to the swamp's surface because of the water depth and soft bottom. An elongate, huge thrashing back, covered with lumpy, bony projections, emerges from the swamp, extending 8 m from the tip of the tail to the front of the jaws. Short, powerful limbs steady the huge animal on the marsh bottom, and its enormous, laterally flattened tail thrashes back and forth to provide more force for the attack and to enhance the bite forces. The jaws are studded with pointed front teeth, 8 cm long, which pierce and hold the theropod's flanks. But

Figure 1.2. *Eastern* Deinosuchus rugosus *attack on a juvenile tyrannosauroid theropod in an Alabama coastal swamp. The crocodylian is represented as an 8.0-m individual. The hadrosaur in the background is a* Lophorhothon atopus. *Drawing by Ron Hirzel.*

the real damage to the prey is inflicted by blunt, low crowned rear teeth, enormously thick and 3 cm wide, nearly solid enamel and dentine for two thirds of their diameter, located close to the maximum leverage region of the jaw hinge where the jaw-closing muscles are at their most powerful. These teeth originally evolved to punch through the hard shells of sea turtles, but they have now been adapted to killing larger and softer prey, and they crush and mutilate the vertebrae of the theropod's lower back, paralyzing its legs. The long jaws open and close a few times, each chomp damaging more bone and nerve tissue. The jaws are able to exert crushing pressures I estimate at more than 18,000 newtons, more powerful than any animal in the modern world—and also more powerful than any *Tyrannosaurus* (Erickson and Olson 1996), probably the most famous extinct predatory giant with massive jaws and teeth.

The stunned, mortally wounded carnivorous dinosaur stands for a moment in shock, unable to breathe because its liver has been damaged, jaws in rictus, and legs quivering from lack of motor control due to severed nerves; and soon it is fortunate enough to die from hemorrhage and the gross damage to its nervous system. It collapses onto its side, into the marsh, and for a few moments, the corpse floats on the water surface. Then the huge predator clamps its jaws on the theropod's legs, shakes its head, rolls its 2.3-ton body, and begins to worry the carcass, tearing off lower limbs and half the tail. After about 10 minutes of this mutilation, a nearly limbless, stub-tailed trunk remains, forming a more compact mass that is pulled into an area of the salt marsh full of tree debris. The larger predator drags the smaller predator's corpse into the snags and impales the remains underwater: it will wait until decay helps to soften the carcass and explode the gut cavity with escaping gases. Then it will completely dismember the torso and consume the entire young carnivorous dinosaur—bones, teeth, guts, offal, and all.

It is the Late Cretaceous, the last epoch of the Age of Dinosaurs—more precisely, around 75 million years ago. The events described were a commonplace attack by a giant crocodylian of the genus *Deinosuchus*. Events like the one I described really occurred and are amply documented by fossil bones, a variety of traces, and inference from the fossil record. Such events occurred not only in Alabama, but also in all southern and mid-Atlantic coastal areas of Late Cretaceous North America from southwestern Texas to New Jersey. Perhaps the most startling aspect of these encounters is the implication that crocodylians, not dinosaurs, were the top predators in coastal habitats of semitropical North America during the Age of Dinosaurs. And despite the impressive size and weaponry of carnivorous dinosaurs, the basic predatory equipment common to most crocodylians proved to be superior in these coastal habitats. (Note here that the term "crocodylians" will be used throughout the book as a general reference to living and fossil alligators, crocodiles, gharials, and their close relatives.) As we observe in modern Africa, a large Nile crocodile has been documented to attack a lion that chose to cross the wrong river (Cott 1961); so we can infer that was the fate of a carnivorous dinosaur out of its usual land base.

Overview of the Age and Habitat of *Deinosuchus*

As my semifictional narrative suggests, the southeastern margin of North America hosted a unique biota during Late Cretaceous time, with a distinctive predator–prey hierarchy dominated by giant crocodylians rather than theropod dinosaurs. In the times and places where the giant species *Deinosuchus rugosus* was extremely abundant, primarily in the Coastal Plains of southeastern United States, it appears from the fossil record that few, large carnivorous dinosaurs coexisted. Indeed, no remains of large theropods have yet been found in *Deinosuchus*-bearing beds in the east, although smaller theropods are fairly common (Schwimmer et al. 1993; Schwimmer 1997b).

Other areas of the continent hosted *Deinosuchus,* especially southwestern Texas (Colbert and Bird 1954; Rowe et al. 1992) and along the shores of the Interior Seaway in today's Wyoming and Montana (Holland 1909). In these areas (see Fig. 1.3), rather than in the southeast, *Deinosuchus* crocodylians reached their largest sizes, and at least in Texas, they were fairly common. However, only in the low country of the southeastern continental margin, corresponding with a broad band

Figure 1.3. Late Cretaceous North America, with its epicontinental and pericontinental seas. The volcanic island chains and atolls spanning the northern Gulf of Mexico and southern Western Interior Seaway are hypothetical but are based on occurrences of volcanic rocks and probable reef limestones of corresponding age. Base map drawing by Ron Hirzel.

of territory from modern central Mississippi to central North Carolina, did the giant crocodylians reach great population abundances. By use of data and assumptions with the imperfections of the fossil record in mind, it still seems that *Deinosuchus* was more abundant in Late Cretaceous coastal Georgia, say, than are modern large crocodiles in most modern habitats. For example, I believe there was a higher population density of *Deinosuchus* in the Late Cretaceous coast of Georgia than there are of *Crocodylus intermedius* in the Orinoco River in South America, or of *Crocodylus niloticus* in the Congo River system in cen-

tral Africa. And it is especially noteworthy that these Cretaceous crocodylians were animals reaching more than twice the body mass of any
living *Crocodylus* or *Alligator* species.

A major focus of this book, besides the giant crocodylians themselves, will be the unique ecosystems and conditions of these southern
Late Cretaceous coastal habitats that enabled such crocodylian populations to develop and flourish for a significant amount of geological
time. It is important to consider that Earth's entire climate during the
Late Cretaceous was relatively warm and equable compared with the
present—and therefore presumably warm enough for cold-blooded
crocodylians to survive quite far to the boreal regions—and that river
and swamp conditions (prime crocodylian habitat) existed all across
the North American continent. As I will detail in Chapter 5, when we
reconstruct the ancient geography and topography of the southern and
eastern coastal regions, we see an especially great abundance of good
crocodylian habitat. But even more significant to the appearance of the
masses of *Deinosuchus* was the influence of the pericontinental seas,
which are defined as ocean waters lapping well over the continental
margins. These onlapping seas, which exist in few areas today, at times
during the past 100 million years extended up and over most of the
southeastern coast. With each transgression, the pericontinental sea
created multiple nearshore habitats in which *Deinosuchus* and other
crocodylians thrived, and these conditions persisted right until the Ice
Age. Fortunately for those interested in the paleontology of *Deinosuchus,* many of these perimarine habitats of Late Cretaceous age (see
Appendix A) are well preserved in the rock record, and fortunately for
my work, some of the best are located within 80 km of my university,
in southwestern Georgia.

Nowhere on Earth today can we find the precise combination of
warm temperatures with continuous, broad areas of nearshore and
marshy conditions to match what the rock record tells us was present
on the southeastern Coastal Plains during the Late Cretaceous. A few
modern areas give limited insights into the habitats and life of such
places—for example, some broad African river valleys and many Australian estuaries and tidal creeks. But these are imperfect analogies
because they are not nearly as extensive as was the southeastern Coastal
Plain of the Late Cretaceous. The best reconstructions show that there
were at least 1800 km of combined and connected salt marsh and
nearshore shallow water habitat, extending from Mississippi to New
Jersey, and all areas are known to have to supported populations of
Deinosuchus rugosus. At least as much linear mileage of potential
crocodylian habitat also existed on the western side of the Interior
Seaway (Fig. 1.3), but it was not likely to have been a continuous band,
as in the east. The region between northern Texas and Wyoming has
produced few Late Cretaceous crocodylian fossils, at least in deposits
that formed close to the sea (there are many freshwater crocodylian
fossils from this region—see Chapter 7). From this and other evidence,
it appears that the terrain along this part of the western shore was
somewhat rugged, not as low as in the east, and the salt marsh areas
were more limited.

Therefore, for these geographic reasons, and for other reasons that will be clear later, the animal life of today does not adequately recapitulate the fauna of the Late Cretaceous Coastal Plains: in terms familiar to those who study historical geology, these would be termed "non-Uniformitarian" conditions. For example, modern life and environments cannot show us many of the biotic interactions that shaped the food webs of coastal Mesozoic ecosystems because, for example, there are no predators alive in North America nearly as large as the theropod dinosaurs *Gorgosaurus, Daspletosaurus, Albertosaurus,* and, of course, *Tyrannosaurus,* all of which lived on the continent during the Late Cretaceous Epoch. In a similar vein, very large herbivores (e.g., the size of rhinoceros or elephants) disappeared from North America at the end of the Pleistocene Ice Age. And even when they were here, they did not habitually venture into the sea or salt marshes of the North American coasts, nor were they likely as abundant as were the ornithopod dinosaurs of the Late Cretaceous. More central to the story at hand is the fact that modern crocodiles do not exceed 6.5 m in length (and may not even reach that size; see Chapter 3). All of these animals and body plans played crucial roles in the Late Cretaceous coastal ecosystems, and we must therefore turn to the deeper fossil record for our insights because the modern world is not very illuminating.

The Late Cretaceous Coastal Plains

Fossils of *Deinosuchus* are quite common in fossil-bearing sedimentary deposits along strips of outcropping rock in eastern, southern, and parts of western North America. These were areas that were coastal during a specific 10-million-year span of Late Cretaceous time termed the "Campanian Age" (see discussion of detailed geological age terminology, Appendix A), but that today are inland by several hundred kilometers from present shorelines (Fig. 1.4). This follows because modern sea levels are approximately 100 m lower than those of the Late Cretaceous, due in part to the standing masses of glacial ice still present on the modern Earth. This ice represents the uptake of seawater; and as far as the record shows, there were no significant glaciers anywhere during the Late Cretaceous. As the sea lowers by loss to standing ice, naturally, the shoreline recedes. The Late Cretaceous was, in fact, the time of the last great epicontinental seas in North America (Hay et al. 1993; and see Chapter 6), which certainly had a major effect on the success of the giant crocodylians.

In the eastern continent, in addition to lowered sea levels, the Appalachian Mountains were heavily eroding throughout the Mesozoic Era. (The same mountain region has been eroding through the Cenozoic Era and continues to do so today, but it does so at lesser rates from when they were newer and higher in earlier times closer to their initial formation during the Paleozoic Era.) Appalachian erosion sent masses of sediment down eastern rivers to the Atlantic and Gulf coasts. The river deposits built up new land surfaces in a variety of depositional settings, all of which contributed to the bulk of land accreting above sea level, at the expense of the shallow, submerged marine continental

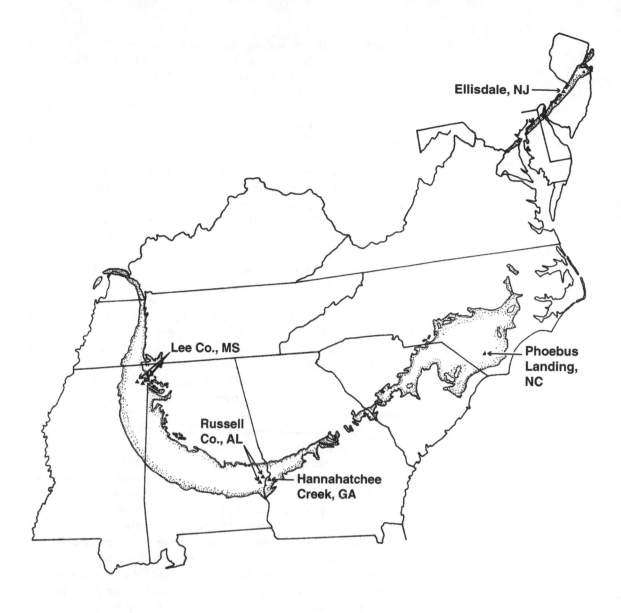

Figure 1.4. Composite eastern
United States Late Cretaceous
rock outcrop, with the locations
of significant Deinosuchus sites
discussed in text.

shelves. The built-up land surfaces included broad alluvial plains, with wide, low rivers, and back swamps, oxbows, and sandbars. As geological time progressed, the courses of rivers meandered wildly, and thus the areas covered by their deposits widened. Where Appalachian rivers emptied to the ocean, some deltas formed, and over time, they pushed the river mouths seaward. The Mississippi Delta comes immediately to mind as a model of this process, and it is indeed being formed in part (i.e., by rivers from the east) by erosion products ultimately carried from Appalachian headwaters. Numerous long-extinct deltas are evident in the sand and silt deposits buried beneath the eastern Coastal Plains, with their ancient outlines and drainages obscured by weathering and overlying younger deposits.

Where river mouths reached the sea, and for a variety of tectonic reasons became deeper and wider, estuaries developed. These provided ideal habitats for the broadest possible varieties of animal life, just as they do today in such places as Chesapeake Bay. (However, in the case of Chesapeake Bay, the actual vast extent of the estuary may have been carved initially by a meteoric impact [Poag 1997], which does not seem to be the case for any eastern Late Cretaceous estuaries we can document.) Over time, estuaries also migrate seaward as the headwaters of rivers aggrade (i.e., as they build up their bottoms). Estuarine deposits are a major constituent of the prehistoric eastern Coastal Plains, but they are not as easily identified as are deltaic deposits. This is because the sediments composing estuarine deposits—dark, muddy sands—are nearly identical to those forming in other nearshore environments, especially back-barrier lagoons (i.e., the marine waters shoreward of barrier islands, which today line much of the eastern and Gulf Coasts and almost certainly did so back through the Cretaceous Period). The geological criteria for recognizing estuarine deposits, as compared with those forming in back-barrier environments, are best determined by examining the fossils because estuarine deposits should contain some brackish and freshwater animals. However, the fossil record is rarely so complete and precise that we may confidently postdict (i.e., predict in the past) that a given fossil assemblage formed in place. That is, at a given fossil locality, we may not be certain whether or not freshwater animals were washed to sea by streams to be deposited with the marine fossils (as in a typical shore deposit), or if those freshwater animals lived in nearby upstream waters (as in an estuary).

Ultimately, a coast is nourished and built up by the sand reaching the shore, to be redistributed by waves, currents, and tides. This coastal sedimentation has been building up the eastern shorelines and extending the landmass of the continent seaward since the early part of the Mesozoic Era, when the Atlantic Ocean and the Gulf of Mexico first opened up to receive sediment. The precise contributions of sea level and sedimentation to long-term marine regression are not easily resolved, but at least we know why the Late Cretaceous coast lies today far from the modern coast. It should be noted that recent observations of the sea encroaching onto land in North America (and globally) is probably an anthropogenic effect of global warming, causing the oceans to warm and swell and glacial ice to melt. This current rise in sea level is a short-term phenomenon of perhaps 150 years' duration. The regression of the eastern coastline from its high point of the Jurassic and Cretaceous to today's beach line was a 100-million-year process with many fits and starts and with high- and low-water times; therefore, a 150-year set of changes in sea level is an insignificant bump on the long-term curve.

In addition to lying far inland, the area across which *Deinosuchus* fossils are found would seem to be much broader than modern shorelines in terms of its onshore–offshore extent. After all, a true shoreline, as the term implies, is a one-dimensional linear feature, whereas an area is two-dimensional. The best analog and understanding of these areally extensive shorelines of the Late Cretaceous comes from modern Okefenokee Swamp in south Georgia (Fig. 1.5). This vast wetland, coinci-

dentally famous as prime alligator habitat, developed during the later Tertiary Period (approximately 5 million years ago) because subtle elevation of the land surface effectively stopped the regional flow of the Suwannee River, which had crossed the region before the land uplift. The upraised, horizontal terrace under Okefenokee resulted in river waters spreading over the landscape, rather than draining through it, creating a vast area of standing water. Similarly, the eastern and southern continental coasts during the Late Cretaceous, which were growing seaward at the time by active sedimentary accretion, developed as nearly horizontal land surfaces. When the seas lapped over the shoreline, they reached far inland across the level surfaces, leaving extensive salt swamps far up shore and creating salt-marsh habitats interfacing with Okefenokee-type freshwater habitats extending far into the land. All of these freshwater and saltwater swamps must have been crocodylian havens, and conversely, they were unusually hazardous for animals less aquatically adapted, such as the dinosaurs of the times.

Even the global position of North America during the Late Cretaceous positively affected the suitability of some habitats for giant crocodylians. Following the breakup of the supercontinent of Pangaea during the later Triassic and early Jurassic, the North American plate began a generally westward-to-northwestward motion as the new Atlantic

Figure 1.5. Okefenokee Swamp, southeast Georgia. This site in the southwestern park region is characterized by cypress groves, blackwater bayous, and dense surface plant growth. Habitats vary throughout the vast wetland region referred to as Okefenokee, and most support large populations of alligators. Photograph by D.R.S.

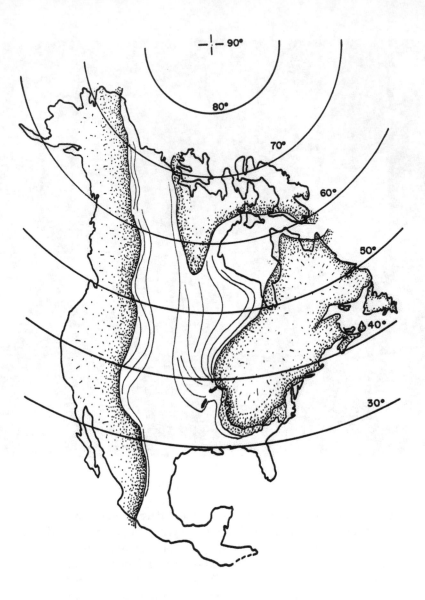

Figure 1.6. Global position of
North America during the
Campanian Age of the Late
Cretaceous. Note that the eastern
Coastal Plain is rotated relatively
southward, and the western side
of the Western Interior Seaway
rotated relatively north–south
as compared with present continental position. Drawing by
D.R.S.; data on continental
position from Scotese et al.
(1988).

Ocean opened. Because the South Atlantic opened sooner, and at a slightly faster rate, than the North Atlantic, the relative position of North America by the Late Cretaceous was more clockwise relative to the present (see Fig. 1.6). And as a consequence of this continental rotation, the position of the eastern Coastal Plains from (present-day) New Jersey to Mississippi was shifted to a more east–west (rather than north–south) orientation. In other words, North Carolina lay largely to the east of Alabama rather than to the north, and most of the eastern Late Cretaceous sites that now contain *Deinosuchus* fossils were rotated southward and probably lay at latitudes less than 35°N.

However, just as the eastern Coastal Plains were located differ-
ently, so were those on the western side of the Interior Seaway. The same clockwise continental rotation caused the Cordilleran (i.e., western) Mountains to be oriented relatively more north–south than at

present, producing more extreme latitudinal change from Texas to Alaska than one observes today. Southwest Texas during the late Cretaceous lay at a tropical latitude around 30°N; however, Montana was pushed close to latitude 50°N. It is interesting to note that despite the higher latitude, fossil crocodylians are still fairly common in the Late Cretaceous of Montana. These fossils in Montana are mostly freshwater crocodylian species, but they also include a few remains of *Deinosuchus,* showing that even with a more northerly position, the overall global climate was still within their temperature tolerance range (Markwick 1998a, 1998b). Nevertheless, crocodylians in general, and *Deinosuchus rugosus* specifically, were inordinately common in the sites located relatively more to the south during the Late Cretaceous (including areas located as far north today as New Jersey). In Chapter 8, I will discuss evidence about the feeding preferences of *Deinosuchus* and speculate that their size and success can be attributed in some part to feeding on dinosaurs. I carry speculation to an extreme and suggest that the presence of any *Deinosuchus* in such remotely northern areas as Montana may have more to do with the availability of abundant dinosaur prey there, as the fossils show, than with the suitability of the general habitat for huge crocodylians.

Introducing *Deinosuchus*

The protagonist of this book has already been introduced as a huge, fierce predator, which it indeed was. But it was also a fairly conventional eusuchian crocodylian—that is, it is among the derived lineages that include all modern crocodylian species (discussed in Chapter 7), and to the casual observer, a *Deinosuchus* would appear to be a cross between a huge alligator and a huge crocodile. On more detailed scrutiny, the observer would notice that the teeth are very thick and unusually blunt in the mid and rear jaws, and that the bony plates (osteoderms) covering the neck, back, and anterior part of the tail were very lumpy and relatively larger than in modern crocodylians. The elongate front portion of the skull, the rostrum, is broad, as in alligators, but slightly bulbous at the tip and with a significant constriction where the fourth lower jaw tooth emerged and crossed the upper jaw when the mouth was closed. This exposed lower tooth is common in modern crocodiles but absent in all living alligators and caimans; however, it turns out to be an ancestral characteristic of all eusuchians. Nevertheless, aside from the mixed broad and constricted aspects of the snout and details of the teeth and osteoderms, the gross appearance of *Deinosuchus* is not much different from an alligator or a crocodile. But of course, there is the aspect of size to consider, with smaller mature animals about 8.0 m long and weighing more than 2 tons. The largest *Deinosuchus* were at least 11.0 m in length and likely reached 12.0 m, with weights over 6 tons—and perhaps much more! These were dinosaur-sized crocodylians in the Age of Dinosaurs.

Most of the knowledge of *Deinosuchus* comes from skulls, jaws, and miscellaneous postcranial bones (i.e., behind the skull). We still do not have well-preserved remains of any *Deinosuchus* that represents a

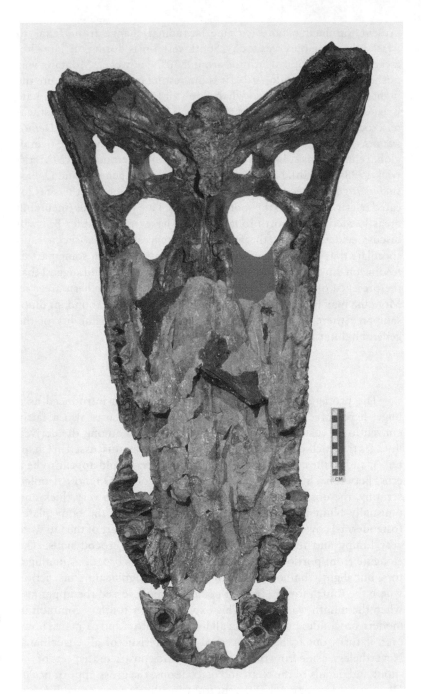

Figure 1.7. Computer reconstruction of the ventral aspect of the skull of Deinosuchus rugosus. *Refer to Chapter 5 for discussions of the specimens used in this reconstruction and Figure 1.8. The posterior portion is the Tupelo, Mississippi, specimen, from the Coffee Sand; the anterior portion is the Lowndes County Alabama specimen from the Mooreville Formation. Both photographs by D.R.S.; computer composite illustration by Jon Haney.*

substantial part of the body. However, the general body plan of eusuchian crocodylians was fully evolved by the time *Deinosuchus* appeared in the Campanian Age of the Late Cretaceous (see Chapter 4 on dating), and it is very well known to us through modern species. All of the postcranial bones that we can reliably attribute to *Deinosuchus* fit

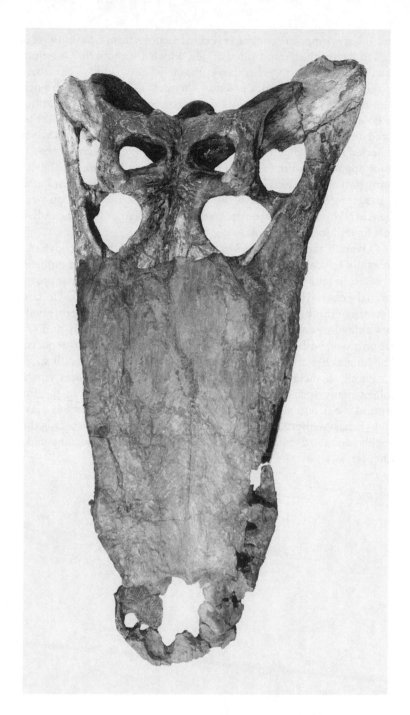

Figure 1.8. Computer reconstruction of the dorsal side of the specimens as described for Figure 1.7. Photograph of the Mississippi skull (posterior portion) by Wann Langston Jr.; anterior unit, photograph by D.R.S.; computer composite illustration by Jon Haney.

nicely into the general makeup of a typical eusuchian, and I am confident that their actual overall appearance is close to the general assumptions that are based on reconstructions. Most of the unique characters that allow the identification of *Deinosuchus* fossils, aside from the size, are in the head, and of these, the teeth are the best. Figures 1.7

and 1.8 are composite computer reconstructions from two skulls recovered in the southeastern United States, which are about 90% complete as a composite (i.e., together they contain 90% of a skull). The histories of these specimens are discussed in Chapter 5, but it is worth noting here that these reconstructions differ considerably from the famous "*Phobosuchus*" skull reconstruction (see Figs. 2.1, 2.5, 2.6, 2.8) featured on display at the American Museum of Natural History (AMNH), collected in Big Bend, Texas. As new materials are collected, it has become apparent that the AMNH reconstruction, which was the basis of many past assumptions about *Deinosuchus*, was inaccurate and greatly exaggerated. Nevertheless, the animal that left the fossils on which the AMNH skull was reconstructed was much larger than the individuals figured in 1.7 and 1.8.

Overall, then, *Deinosuchus* was an incredibly big crocodylian of generally typical modern proportions and appearance. Its uniqueness stems from both the size and its ecological role as the dominant predator in the coastal habitats of Late Cretaceous North America. The cover artwork of this book, by D. W. Miller, is the most accurate portrayal of this animal to date and features anatomically correct body and skull proportions, with special attention given to the teeth and osteoderms. The animal life around the crocodylian in the painting are all species commonly associated with *Deinosuchus* fossils in the sites where I collect many specimens, and the theropod leg remains in the foreground is a tribute to the predatory role I assume *Deinosuchus* played as an occasional predator of predators. This predatory aspect is thoroughly discussed (might one say chewed over?) and documented in Chapter 8.

2. The Early Paleontology of *Deinosuchus*

Taxonomic Conundrums

According to rules governing the classification of both recent and fossil animals, the scientific name that is given in the first description of a particular organism becomes permanent. Thus, the original species description and the actual specimen or specimens figured and specified in the first scientific publication—termed the type specimens—become the immutable name references. There are means to take exception with the rules of classification under extraordinary circumstances, but most taxonomists avoid the required legalistic procedures and follow tradition. These conventional taxonomic practices have often caused confusion and dismay in popular culture when they involve popular creatures. One of the most famous examples concerns the invalid sauropod dinosaur name "*Brontosaurus*," which had to be replaced after many decades of common use (notwithstanding a fairly recent U.S. postage stamp) because an obscure description of scanty material named *Apatosaurus* was published before the famous *Brontosaurus* skeleton was described (see Gould 1992 for a good discussion).

Taxonomic protocol has also caused many problems and disagreements among professional paleontologists in more technical contexts, especially when a poor specimen has been the legal name-bearer, while a good fossil becomes widely recognized as the organism's identity. The case of the huge Late Cretaceous–age teleost fish, commonly called "*Portheus molossus*," comes to mind. A beautifully preserved large specimen was reported by Edward D. Cope under this name in 1871 from the Smoky Hill Chalk in western Kansas. This Late Cretaceous fish is actually quite common as fossils (and, coincidentally, a similar species co-occurs in some deposits in the eastern United States with *Deinosuchus rugosus*) and can be found in many natural history museums,

Figure 2.1. Big Bend skull reconstruction, formerly at the AMNH. Computer composite illustration by Jon Haney.

some still labeled "*Portheus molossus.*" But, sadly for simplicity's sake, a single pectoral fin spine of the same fish was described just a single year earlier (1870) by Joseph Leidy and named *Xiphactinus audax.* Therefore, Leidy's is the correct and permanent name for the common Late Cretaceous "bulldog fish" in our museums, despite its poor representation by the legal type specimen.

Deinosuchus rugosus, both as a genus and species, suffers from a similar problem, but perhaps to a lesser degree than the animals just cited. The general public, and most paleontologists, became aware of the creature after the spectacular Big Bend specimen was prepared, reconstructed, and displayed in the American Museum of Natural History (AMNH) in New York City—under the name *Phobosuchus riograndensis* (Fig. 2.1). The generic assignment was formally corrected to *Deinosuchus* in 1979, leaving only the species designation still in current contention. However, the "*Phobosuchus*" skull reconstruction, which is by far the most famous specimen of *Deinosuchus* to date, is neither the name-bearer for the genus nor for the species. It remains among the most important of *Deinosuchus* fossils, from what is certainly the most important of localities in which the crocodylians occur. Nevertheless, the generic name "*Phobosuchus*" was never valid, and for arcane reasons of taxonomic protocol, the specific (i.e., species) name "*riograndensis*" must also be suppressed in favor of two obscure teeth originally named *Polyptychodon rugosus,* as will be explained in this chapter.

As of this writing, a formal, technical publication evaluating the taxonomic relationships of all *Deinosuchus* materials in the United States has not yet been published. I hope that this will be done soon, although its publication might make the following discussion in this book obsolete. But in the meantime, both in deference to taxonomic procedure and to keep names and sources of information clear, whenever I refer to a type specimen of a deinosuchid, I will refer to it by the original name given in the first description. I will also refer to topotype material (i.e., that from essentially the same locality and stratum) by the original name. Because I am writing this book with the knowledge that all deinosuchids are now considered to be congeneric (i.e., assigned to the same genus), and working under the assumption that they are also conspecific (i.e., of the same species), some organized method regarding name usages needs to be in place to indicate these understandings. Therefore, where I refer to a specimen that formerly went by a name we should change to *Deinosuchus rugosus* (except when that specimen was first described), I will put the changed portion of the name in quotation marks and leave any unchanged portion as is. Thus, the reader will see frequent references to *"Phobosuchus" riograndensis* from the Big Bend region of Texas, and Montana or Wyoming specimens of *Deinosuchus "hatcheri,"* as well as a few references to *"Polyptychodon" rugosus.* All of these are original type designations of the materials, and some are still in use by a few workers in the professional literature.

The Big Bend Monster

Big Bend National Park covers a huge area, occupying (as the name indicates) the north side of a large bend of the Rio Grande River (the Rio Bravo to Mexicans) in Brewster County, southwest Texas (Fig. 2.2). It is one of the most remote of the national parks of the United States, located more than 500 km from the nearest major airport in El Paso, with plenty of desert wilderness still available to escape from tourist crowding and with plenty of vertebrate fossils visible (but not legally collectable) to the observer. Even today, access to some areas in the park can be difficult and hazardous. The climate is profoundly hot and dry in summer, with potable water scarce, and poisonous, biting, sticking, and stinging organisms abundant. During the 1930s, when exploration for fossils in the park began in earnest, it was quite an adventure to prospect the region because of the physical hardships. However, for geologists and paleontologists (and perhaps other categories of sociopaths), the park and surrounding areas offer opportunities for exciting discoveries because the Big Bend region broadly exposes deposits of many geologic ages, with the Cretaceous and Tertiary Periods especially well represented by fossiliferous sediments. There is a palpable sense of adventure when one searches for Cretaceous fossils in the several formations exposed in Big Bend because new discoveries are always possible, even expected.

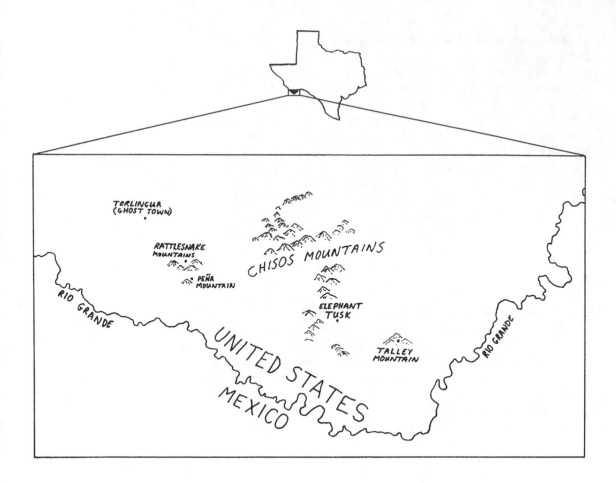

Figure 2.2. Locality map of selected sites in the Big Bend region, with inset showing the position of Big Bend National Park in North America. Drawing by Ron Hirzel.

The Cretaceous deposits of Big Bend will be the focus here, including units deposited in both marine and nonmarine environments. Most notable for our purposes is the Aguja Formation, a mixed marine and nonmarine rock unit from which the important *Deinosuchus* remains come. This formation has six members (i.e., formal subdivisions), some of which are interpreted to include beds containing coastal marsh and delta sediments, with fossils of freshwater turtles and fish, smaller crocodylians, and dinosaurs. But there are also marine beds in the Aguja, containing fossil shark teeth, oysters, marine turtles, and *Deinosuchus* (Rowe et al. 1992). Such multienvironmental stratigraphic deposits as the Aguja Formation are useful for paleontologists interested in paleoecology. By their multiple origins, they allow us to trace relationships between the dinosaur-bearing nonmarine deposits, common on the western side of the Interior Seaway in the western United States (see Chapter 5), with the largely marine deposits in the eastern United States. For example, the shark species *Scapanorhynchus texanus* and *Squalicorax kaupi,* as well as *Deinosuchus rugosus* and several mollusks, which are all present in the marine Aguja beds in Texas, match one for one with contemporary strata in the eastern Coastal Plains from North Carolina to Mississippi. This similarity of marine fossils contrasts with the dinosaurs in the two regions, which are quite different. For example, there are no Late Cretaceous ceratopsids (horned dinosaurs) known in the eastern United States, whereas they are very common on the western side of the Interior Seaway. Also, the relationships between the larger hadrosaurs (duck-billed dinosaurs) found in east and west are unclear (Lamb 1998; Schwimmer 1997b), as are relationships among the larger carnivorous dinosaurs. Thus, we can readily correlate the marine beds across the interior of the continent, but it is not nearly as clear for the nonmarine beds.

Reports and rumors of fossil bones in the Big Bend region had begun to arrive in the late 1800s, and some of these may even have concerned *Deinosuchus* (Langston et al. 1989). However, detailed search and study of the park's vertebrate fossils really began with federal government projects of the mid-1930s. During the Great Depression, many people were employed under the auspices of WPA (Works Progress Administration) and the CCC (Civilian Conservation Corps) to help alleviate some of the job shortages. Among the projects funded by the agencies was the development of fossil quarries in National Parks, including the (then) proposed Big Bend National Park. Several WPA quarries were dug by local unskilled workers in the late years of the 1930s, and those of interest here are quarries developed within the Late Cretaceous Aguja Formation, northwest of Talley Mountain, in the general Chisos Mountain region of the park (Fig. 2.2). During the quarrying operations, the sites were visited by personnel from several Texas and Oklahoma universities (Davies and Lehman 1989), but little hard-edged paleontology was done during the period of active quarry operations. The original purpose of these quarries was to produce at least one respectable-looking, mountable dinosaur skeleton for display in Texas—a goal that was never accomplished. However, as a by-product of the operations, from a small exposure of the Aguja Formation, less than a ½-km distance from one of the abandoned WPA quarries (Fig. 2.3), came the spectacular pieces of the best-known deinosuchid skull—and the origin of the *Deinosuchus* mystique.

Edwin H. Colbert and Roland T. Bird (1954) reported the specimen in an *American Museum Novitates* article, noting it as "one of the principal discoveries of the 1940 paleontological expedition of the American Museum of Natural History (AMNH) to the Big Bend Region," which is technically true, but which doesn't reveal that it was Texas institutions and the WPA that initiated the work leading to the discovery (Langston et al. 1989). The Colbert and Bird paper named the animal "*Phobosuchus riograndensis*" by using a generic name to which the German paleontologist Franz Nopcsa (1924) had recently reassigned several giant crocodylian groups, and a species name that was derived from its occurrence near the Rio Grande River.

Roland T. Bird, together with Barnum Brown, was the link between the original dinosaur exploration in Big Bend and the "*Phobosuchus*" discovery. Bird was the hyperenergetic field operative and preparator of the AMNH who, together with Barnum Brown, curator of vertebrate paleontology at AMNH, helped to make the AMNH holdings the most complete Mesozoic vertebrate fossil collection in the

Figure 2.3. Talley Mountain site in Big Bend National Park, close to the locality of the original "Phobosuchus" riograndensis discovery. The oddly shaped peak in the left background is called "Elephant Tusk."

world (then and, arguably, now). E. H. Colbert, the primary author of the study, was the renowned senior vertebrate paleontologist at AMNH for three decades and author of a leading basic vertebrate paleontology textbook, as well as dozens of popular books on dinosaurs and other vertebrate life. Indeed, an earlier edition of Colbert's vertebrate paleontology textbook was the one I read for my first course in the subject.

According to R. T. Bird's own writing (Bird 1985, edited after his death by V. T. Schreiber), Bird was collecting Early Cretaceous dinosaur tracks in the Paluxy River beds in central Texas when the big crocodylian specimen was located and collected by Brown and field assistant Erich Schlaikjer. Bird and Brown, shortly after the discovery, visited the "hole in the hill" on the flanks of Talley Mountain from which it came, and later Bird was assigned the responsibility to prepare the material when it was shipped back to the AMNH. Bird was daunted by the poor quality of the bone, encased in thick overgrowths of unusually resistant limestone. But he also observed and was intrigued that one of the teeth poking out of the plaster jacket, obviously crocodylian in form, was double the size of any he had ever seen. Indeed, the tooth he observed was larger than those of the famous *Tyrannosaurus rex* skeleton mounted one floor down from his preparation lab at the AMNH.

The *"Phobosuchus"* specimen was prepared gradually, but the quality of the bone surfaces was not good. As preparation proceeded, Bird and Brown (who was observing the preparation) realized that their most significant discovery at hand was of a set of astoundingly huge, bulbous premaxillae (the paired bones of the snout). The proportions of these bones were so unusual (Figs. 2.5, 2.6, 2.8) that Bird stated he originally believed they were the ilia (parts of the hips), figuring that what turned out to be the tooth sockets were actually the acetabula (where the heads of the femora would articulate, if they were ilia). After examining comparable modern crocodile skulls—especially, by Bird's account, that of very large *Crocodylus porosus* (Fig. 2.5)—it became quite obvious that these oddly shaped premaxillaries, although recognizable and identifiable, were much larger and different from those of any modern crocodile.

The remaining materials in the plaster jackets brought back from Big Bend contained a disappointing assemblage of additional bones (Colbert and Bird 1954), nearly all from the skull (refer to Fig. 3.1 and

Figure 2.4. (left) Dentary fragment and (right) dorsal vertebra from the "Phobosuchus" riograndensis holotype specimen from Big Bend. Note the unerupted tooth in the much larger alveolus in the dentary. Scale units are inches and centimeters.

Appendix B for a glossary containing the following anatomical terms). There was a substantial portion of the anterior dentary, part of the right anterior maxilla, a fragment of the left surangular, and seven teeth. Representing the postcranial skeleton was a single vertebra from the anterior thorax (bearing transverse processes for rib articulations); part of the right scapula; a possible right ilium, badly distorted and incomplete; and a few osteoderms (dorsal dermal ossicles, as discussed in the next section).

R. T. Bird, with help from many AMNH personnel and with help from comparisons with modern and subfossil crocodile skulls, reconstructed the Big Bend skull combining the real fossil bone—such as there was—with a lot of plaster (Fig. 2.6). It was really a beautiful job of artistic sculpting. Colbert and Bird (1954) stated that the reconstruction was based on the proportions of a medium-sized Cuban crocodile, *Crocodylus rhombifer* (Fig. 2.7), although Bird's memoirs recall that he used a very large *Crocodylus porosus* as the model. When completed, this specimen was on display at the AMNH for more than 40 years, around the corner from the Hall of Cretaceous Dinosaurs. I can remember the fear and awe a four-year-old felt staring at the 2-m skull in the corner glass-walled mount, with larger-than-life jaws propped open

wide and a throat more than sufficiently large to swallow even an overweight child. I remember hoping that the animal was really very dead and that the glass was very thick. The background diorama of a painted scene of Late Cretaceous Texas, with a small ceratopsid dinosaur (probably a *Centrosaurus*) being threatened by a rampant "*Phobosuchus*," reconstructed the body of the crocodylian to look very much like a heavy-bodied Australian or Nile crocodile.

Aside from the huge size, the AMNH skull reconstruction featured several anomalies. The premaxillae have large fenestrations (i.e., openings—literally translated as "windows") in their anterodorsal surfaces, connecting with large internal hollows in the snout. These fenestrae almost look like nostrils, but they are not because the real nostrils are evident on the dorsal surface. In fact, these extra holes may have resulted from Bird's difficulty in extracting the specimen from the limestone matrix, given the possibility that he removed bone along with the matrix over the thin areas at the tip of the snout. Colbert and Bird (1954) considered these premaxillary fenestrae to be unique diagnostic characters (technically, "autapomorphies") of "*Phobosuchus*" riograndensis, but to date, no additional specimens have been collected with this region preserved sufficiently well to determine whether or not

Figure 2.6. AMNH "Phobosuchus" skull reconstruction, left lateral view, with Roland T. Bird holding the ruler. Reproduced with permission, negative 318633A, photograph by Charles H. Coles (1942), courtesy of the Department of Library Services, AMNH.

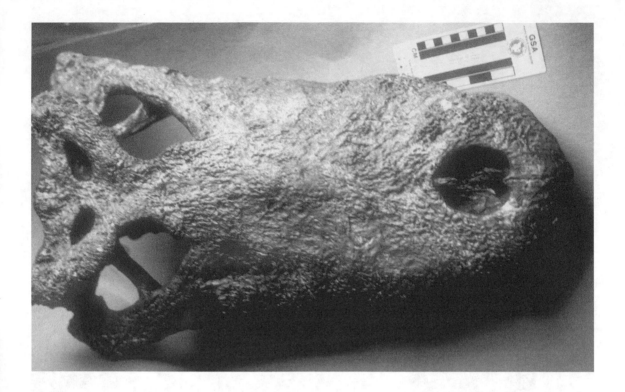

Figure 2.7. Large, subfossil Crocodylus rhombifer *skull from AMNH collections, possibly the specimen that served as the reconstruction prototype for the "Phobosuchus" reconstruction.*

they are a real anatomical feature. The premaxillae were also reconstructed with exaggerated flexure where they meet the nasal and maxillary bones (Fig. 2.6), forcing the front teeth to point straight down ventrally and the notch between the premaxillae and maxillae to be exaggerated. This extreme bending also forced the reconstruction of the snout to bulge upward strongly (and unnaturally).

In addition to the problems cited above, the AMNH skull reconstruction tapers too strongly from back to front in dorsal aspect (Fig. 2.8). Because the actual bone recovered from Big Bend gave no control on the proportions of the posterior skull width, the overall skull shape was entirely conjectural—and ultimately proved to be incorrect. We now know from additional specimens that the premaxillary region of *Deinosuchus* is proportionately broader than in modern true crocodiles (i.e., *Crocodylus* species), and in fact, this may be attributable to its closer affinities to the broader-snouted alligators rather than with crocodiles (see Chapter 7). Not knowing this fact, and given the indisputably enormous size of the snout, the excessive tapering of the Colbert–Bird reconstruction required to fit the huge premaxilla into the crocodile configuration produced an exaggeratedly enormous width across the back of the skull. And compounding the exaggeration, by proportionately modeling the skull height to its width, the error substantially increased the skull's massiveness. Thus, although the Big Bend skull would be indeed huge by any view, the errors in reconstructing its posterior width, height, and mass significantly affected extrapolations

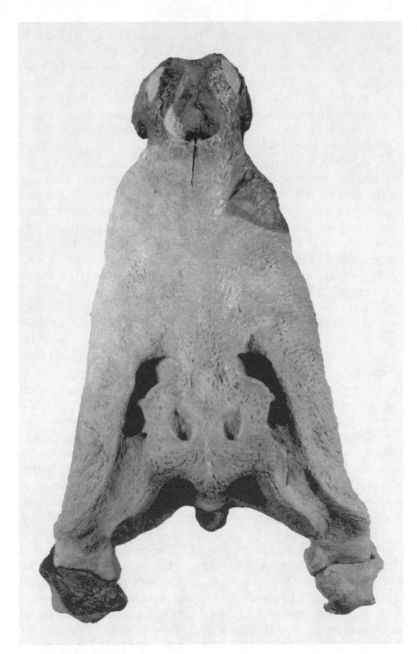

*Figure 2.8. AMNH
"Phobosuchus" skull, dorsal view
taken during reconstruction.
Note that the darker areas are
actual fossil bone, whereas the
majority of the specimen is light-
colored, artificial sculpted
material. Reproduced with
permission, negative 318635,
photograph by Charles H. Coles
(1942), courtesy of the Department of Library Services, AMNH.*

of the animal's overall size. Colbert and Bird (1954) estimated the total
length of the Big Bend specimen at 50 feet (15.2 m), an extrapolation
that may be reasonable from their skull reconstruction but that is much
too large.

Presently, the AMNH "*Phobosuchus*" skull has been taken down
from display and the actual bones have been removed from the reconstruction. Indeed, all of the famous dinosaur displays on the fourth
floor of the AMNH have been renovated and rearranged in the past half

decade, so that the absence of the "giant crocodile skull in the corner glass mount" may not be as noticeable as it might have been during the 40-plus years that it terrified small children. The Big Bend *Deinosuchus* specimens have been under study for many recent years by Wann Langston Jr. at the University of Texas–Austin Vertebrate Paleontology Laboratory. Professor Langston has been working and supervising research on vertebrate fossils of the Big Bend since 1963 (Langston et al. 1989). He and many students and colleagues at the University of Texas have recovered new and important specimens, including a large left *Deinosuchus* mandible, two partial skulls, and three new specimens that at the time of this writing are still undescribed. One of these new specimens is from an animal as large as the AMNH material (W. Langston Jr., personal communication), and when it is restored, it may be the most complete *Deinosuchus* specimen known to date. In addition, Dr. Langston and others have collected many new, unassociated *Deinosuchus* osteoderms, teeth, and other bones in the Aguja Formation (Rowe et al. 1992).

The generic name *Phobosuchus* was used for the Big Bend specimen by Colbert and Bird in 1954 because Franz B. Nopcsa had previously created the new generic name in 1924 in a brief and rather confusing taxonomic paper on fossil Brazilian crocodylians. Nopcsa (1924) attempted to straighten out what he considered some inappropriate junior synonyms (i.e., new names for previously defined taxa), including "*Dinosuchus*," an apparently invalid designation based on a single, huge crocodylian vertebra from South America. Nopcsa also subsumed several established giant crocodylian taxa into the new generic name he coined, *Phobosuchus*. Among the species he included in *Phobosuchus* was one from Montana, *Deinosuchus hatcheri* (note the one-letter spelling difference from *Dinosuchus*) which happens to be the genotype (i.e., the generic name-bearer) of our crocodylian under discussion. Because of this 1924 publication, which was still considered valid as of 1954, the arcane rules of taxonomy required that Colbert and Bird must assign the Big Bend specimen to *Phobosuchus*. On reflection, *Phobosuchus* was not really a bad name for the huge crocodylian because it means "fearsome crocodile." This name persisted for the Big Bend giant crocodylian, and others, until relatively recently. Over the years, it became apparent that the name "*Phobosuchus*" was an unnecessary taxonomic invention for several reasons, including the fact that "*Dinosuchus*" was not "*Deinosuchus*." It became clear that most of the species Nopcsa placed in "*Phobosuchus*" were not evolutionarily related; that is, in modern taxonomic terms, it was a polyphyletic, artificial genus and was thus invalid. In 1979, Donald Baird and John R. Horner assigned some common Late Cretaceous giant crocodylian fossils from North Carolina to *Deinosuchus*, and in the same article, they suggested that the Big Bend specimens also belonged to that genus (discussed later in this chapter). However, in 1979, they did not yet argue that all of the North American deinosuchids were of one genus; thus, "*Phobosuchus*" still existed as a concept to that date.

The Generic Namesake from Montana

Compared with the notoriety of the Big Bend specimen, the name-bearing fossils of the genus *Deinosuchus* have been seen by few non-specialists, and the skeleton has never been reconstructed for public display. W. J. Holland (1909) reported on the specimens, originally collected in 1903 by T. W. Stanton and John Bell Hatcher of the U. S. Geological Survey, in a brief article in the *Annals of the Carnegie Museum,* Pittsburgh (where the specimens are cataloged and housed). Holland reported that some bone fragments were found exposed on the ground surface near Willow Creek, in Fergus County, Montana. Shortly after the original finds, a third person, Mr. W. H. Utterback, was sent to the site, and he found several hundred additional fragments, but no complete bones. Holland (1909) reported that, upon recognizing the animal leaving all these big bone fragments was a huge crocodylian: "Mr. Hatcher immediately lost interest in the material . . . and then came his untimely and melancholy end" (!). We don't know what John Bell Hatcher was expecting to find, but he was among the famous dinosaur bone hunters of the late 19th century, chiefly under the auspices of O. C. Marsh of Yale University. Hatcher accompanied W. J. Holland and many other famous fossil bone hunters in the frenzy of explorations and excavations in the western United States (Colbert 1997) and was responsible for discovery or recovery of some of the best-known American dinosaurs, including *Diplodocus longus* and *Triceratops horridus*. Hatcher and Holland were each authors of important monographs on Jurassic and Cretaceous dinosaurs—and therefore we may assume Hatcher had hoped the Willow Creek specimen was a dinosaur. The "melancholy end" of Hatcher as reported by Holland was due to typhus, at the young age of 42 years (see Dodson 1996). The job of describing the bones was apparently passed on to Holland, at the urging of another leading vertebrate paleontologist of the times, Samuel W. Williston.

None of the bones from the type lot from Fergus County is clearly from the skull or mandibles, nor are any good teeth preserved. What is present and noteworthy among the identifiable remains are two huge vertebrae, a pubic bone, a few ribs, and, especially, 25 large, thick, oddly irregular osteoderms. It is the gigantic size of the vertebrae (and other bones), combined with the distinctive morphology of the osteoderms (Fig. 2.9) that justified Holland's creation of the new generic name for the crocodylian. Holland coined the genus *Deinosuchus* from the roots *deino-* (= terrible) and *suchus* (= crocodile), and he named the species after the late Mr. Hatcher (who died in 1904); thus, we have *Deinosuchus hatcheri*. The size of the vertebrae was carefully detailed by Holland, who noted that these were the largest crocodylian vertebrae ever measured. Their shape demonstrated that the animal was an advanced (eusuchian) crocodylian because the vertebrae are procoelous (Fig. 2.10). That is, in vertebrae where there is a ball-and-socket articu-

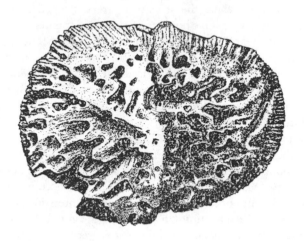

Figure 2.9. Drawings of representative Deinosuchus hatcheri *osteoderms from the Judith River Formation (from Holland 1909). The upper two smaller illustrations are dorsal (left) and lateral (right) views of a small cervical osteoderm. The lower figure is the dorsal view of a large cervical 'derm, approximate width 14.0 cm. Reproduced courtesy of the Carnegie Museum of Natural History, Pittsburgh, Pennsylvania.*

lation between the adjacent centra (the main mass of the vertebra), the socket end (= *coel*, hollow) is forward (= *pro*). Eusuchian crocodylians, which include all living forms, are characterized by having strongly procoelous dorsal vertebrae (i.e., the vertebrae located between the hip and the neck), whereas less derived crocodylians usually had other morpologies among their vertebrae. The most anterior vertebra from the *Deinosuchus* type set comes from the anterior thoracic area, and therefore it has prominent transverse processes (also termed diapophyses or, simply, rib attachments). It has an anterior–posterior length of 14 cm, a width across the centrum of 12.2 cm, and is 68 cm across the transverse processes. In fact, were the shape of this vertebra not obviously that of a eusuchian crocodile, one might assume from the size that it came from a sizable dinosaur! A posterior dorsal vertebra from the same type series is slightly longer (15 cm) but narrower (11 cm) across the centrum. Both of these vertebrae are so huge that estimates of the animal's overall length ranging up to 16 m have been presented in scholarly publications (e.g., Steel 1973). This length is a considerable overestimation, as will be discussed in the next chapter; nevertheless, the *D. hatcheri* individual was an extraordinarily huge animal, possibly a bit larger than the "*Phobosuchus*" *riograndensis* type specimen.

The osteoderms (also termed osteoscutes, scutes, or 'derms) associated with the Montana specimen are unusual and may ultimately be the single most diagnostic feature of *Deinosuchus*. Crocodylians are only one of many groups among the Mesozoic reptiles commonly termed

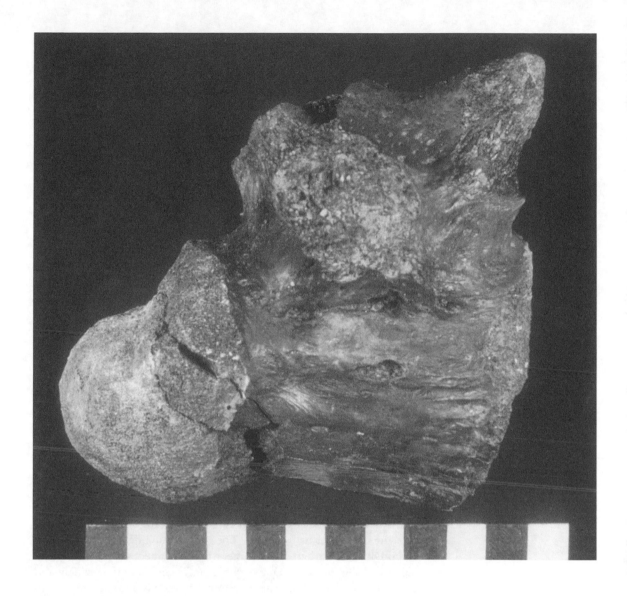

archosaurs (see Chapter 7) that often develop extensive overgrowths of dermal bone. Other examples of bony archosaurs include several groups from the Triassic Period, such as the armored, spiked, herbivorous aetosaurs (including the North American genus *Desmatosuchus*), and the large, predaceous rauisuchids (see p. 142). Many dinosaurs too have dermal bone: obvious examples are the heavily armored ankylosaurs and the stegosaurs with their prominent dorsal bony projections. Less well known is that some gigantic sauropods also have been shown to develop a discontinuous set of osteoderms (McIntosh et al. 1997).

Crocodylian skin ossification is especially consistent in morphology and is usually present over much of the dorsal surface. Exceptions to the usual presence of 'derms in crocodylians are several fully marine families of the Jurassic and Cretaceous (Chapter 7), where the dorsal

Figure 2.10. Lateral view of a typical, larger Deinosuchus rugosus *presacral vertebra from North Carolina. The concave end of the centrum is to the right (anterior); thus, it is procoelous. Scale bars are 1.0 cm each; total length of centrum is 9.7 cm. Specimen courtesy of the North Carolina State Museum.*

dermal bones were secondarily lost, probably because they were unnecessary for support purposes in the water (as discussed below) and would have hindered free up-and-down propulsive movement of the trunk and tail. Osteoderms in crocodylians do not serve primarily as protective structures, as one might assume initially. Frey (1984) demonstrated that they are the structural analog of a box work bridge suspension system—that is, a system of individual load-bearing bones that act collectively as an external tensional system. By explanation, Frey argued that the major groups of "high-walking" land vertebrates—mammals, dinosaurs, and crocodylians—must have a significant muscular bracing system across their dorsal surfaces to give the vertebral column enough strength to form the arch necessary to elevate the body off the ground. In the case of mammals, there are high neural spines on most vertebrae to attach and brace the epaxial muscles (i.e., the dorsal muscles external to the vertebral column). In dinosaurs, the massive tails and pelves serve as the source of leverage for dorsal stresses, in addition to some groups that also have sizable neural spines. By contrast, crocodylians have neither high neural spines nor massive tail and pelvic bones to serve as muscle attachments. However, because the osteoderms are deeply inserted in the dorsal muscles, they apparently work together as the load-bearing reinforcement across the back. This bone and muscle complex can actually exert sufficient upward force to pull the crocodylian's body off the ground surface, keeping the belly from dragging and allowing the feet to be brought somewhat under the body during walking. This biomechanical engineering is analogous to the flying buttresses of Gothic cathedrals, where the support structures are placed externally to the masses needing support to increase the leverage of applied forces. The 'dermal support can enable a very heavy, low-slung animal to move overland. The rapid, terrestrial "high-walk" locomotion of larger crocodylians (Cott 1961) is possible only because of the muscular tension directed across the 'derms and translated across the chest and belly region. The reason osteoderms are pitted in crocodylians is for better attachment of the connective tissues anchoring dorsal skin muscles. One may also infer generally that widely and deeply pitted 'derms imply greater amounts and need for dorsal musculature in a high-stepping crocodylian.

Nearly all vertebrate-fossil collectors can recognize the presence of crocodylians in a fossil assemblage by the appearance of the characteristic osteoderms. However, it takes some experience to differentiate species by details of the 'derms because those of many crocodylian taxa are similarly shaped and pitted. Nevertheless, the presence of *Deinosuchus* in a fossil assemblage can be easily recognized because the osteoderms are distinctive in size, shape, and pitting (Fig. 2.11). And all of the details from *Deinosuchus* 'derms show that they were indeed adapted to help support massive animals on land.

Holland used a specimen of *Crocodylus acutus* (the American crocodile) as his model to analyze the shapes and body positions of various osteoderms. He observed 92 on the living species, and from these, he inferred that with the *Deinosuchus hatcheri* specimen he was observing 'derms from all body regions. These are termed (from ante-

rior to posterior) nuchals, cervicals, dorsals, and sacrocaudals. All of the *Deinosuchus* osteoderms contain wide, deep pits on the dorsal surface; some pits are more than 1 cm wide and are nearly as deep. Many of the pits intersect and create deep grooves, mostly running from center to margin. The osteoderms are also inordinately thick, with some of the smaller-diameter cervical and sacrocaudal specimens approaching a hemispherical shape (Fig. 2.11) because they are nearly as high as they are wide. The 'derms in the huge type individual in many cases show a poorly defined outline, and few of them are really symmetrical. In addition, there are various shapes, sizes, and degrees of keels (vertical ridges) among the type specimen's 'derms, with most being quite vaguely keeled. Among crocodylians, the presence of keels on osteoderms is variable, ranging from some taxa with nearly flat 'derms, totally lacking keels, to some with high, well-defined keels, including variations with split or multiple keels on single 'derms.

    Deinosuchus specimens discovered later, including many isolated osteoderms in Eastern Coastal Plain deposits and from the Aguja Formation in Big Bend, generally have more regularly shaped 'derms than the Montana type specimen. A well-preserved small skeleton from the Coastal Plain of Alabama (see Chapter 5) has many 'derms associated

Figure 2.11. Representative Deinosuchus *osteoderm (dorsal view) from the eastern United States, probably a cervical (cf. Fig. 2.9). The lower margin is slightly ablated. Scale in centimeters. Specimen courtesy of the North Carolina State Museum.*

with it that are even more symmetrical and well defined than are larger specimens, yet they show the wide deep pits that diagnose *Deinosuchus*. All these observations suggest there is a relationship between the overall size of the animal and the regularity of shapes of the osteoderms. This makes sense, given that the osteoderms of larger individuals must be proportionally better able to serve the function of helping to support the body when the crocodile is moving on land in the high-walk posture, with the body suspended off the ground surface. Given that the stresses in larger *Deinosuchus* would be enormous, it makes sense that the osteoderms bearing such stresses would grow massive and less regular in proportion and would have deep pits for strong attachment of skin and connective tissue. It is obvious that larger *Deinosuchus* did not need heavy dermal armor for protection because they were the largest predatory animals in nearly all their environments. The sole exception, perhaps, may be in deeper marine shelf areas, such as western Alabama, where very large mosasaurs and plesiosaurs are found. Even so, plesiosaurs and mosasaurs were almost certainly predators of fish, smaller mosasaurs, and cephalopods and did not have the type of teeth that would suggest they might attack thick-skinned, formidable creatures such as an adult deinosuchid.

It is also reasonable to infer that the presence of such heavy and deeply pitted osteoderms in large *Deinosuchus* prove that the animal could and did walk on land—otherwise, the 'derms would serve no particular purpose and would not likely have evolved such a distinctive morphology. Further, we may be bold enough to speculate that the massive 'derms of the Montana specimen even indicate the possibility of high-walk locomotion in the largest *Deinosuchus* specimens. This is a remarkable possibility because it allows us to envision massive creatures (of sizes discussed in detail in the next chapter) able to move rapidly on land despite short legs, and moving even more efficiently in water by tail propulsion and the advantage of buoyancy.

The type specimen of *Deinosuchus* from Montana is an anomaly in several aspects. It seems to be a rare occurrence of the crocodylian at the fringe of its normal geographic range, that also happens to be among the larger representatives of its taxon; and by historical coincidence, it happens to be the genotype (genus name-bearer) to boot. After the report by Holland (1909), only a few remains of *Deinosuchus* (mostly osteoderms) have ever been found in the western United States located north of Big Bend, Texas. This is especially surprising because other crocodylian remains are relatively common in all western U.S. Cretaceous deposits, from both marine-shore and nonmarine deposits (e.g., Carpenter and Lindsey 1980; Erickson 1976; Horner 1989; Lehman 1997; Markwick 1998a, b). Most of these western crocodylian occurrences are nonmarine, smaller forms, such as species of *Leidyosuchus* and *Brachychampsa* (discussed in Chapter 7). However, it is possible that some unrecognized occurrences of *Deinosuchus* may exist because it is not always possible to determine whether a Late Cretaceous crocodylian fossil came from a smaller *Deinosuchus* or another taxon. This is the case unless the most diagnostic parts (teeth and osteoderms) are preserved, or unless the fossil represents an enormous individual (which would likely be a *Deinosuchus*). Still, overall, it is surprising that *Deinosuchus* was not more abundant along shore areas of the northern part of the western seaway, given that the *D. hatcheri* specimen obviously existed there.

The environmental setting inhabited by the Montana *D. hatcheri* is not known with certainty—and in fact is a matter of contention. The Judith River Formation from which it was reported by Holland contains several units (Horner 1989) representing marshy–terrestrial and nearshore–marine habitats (just as did the Aguja Formation in Big Bend, the strata producing the "*Phobosuchus riograndensis*" type specimen). Any of these habitats could potentially be suitable for crocodylians, and so the precise unit from which came the *D. hatcheri* type specimen is of some importance. It is especially important to have this knowledge when we try to reconstruct the habitat preferences and range of *Deinosuchus* because the genotype specimen is the only one alleged to have come from nonmarine strata. Holland (1909) did not discuss other fossils associated with the type, and the site cannot be located with enough precision to be certain which unit was the host: it may have been from the marine (Bearpaw) lithofacies or from nonmarine or marginal marine Judith River lithofacies. Although we do not know whether the *Deinosuchus* genotype was found in marine or nonmarine deposits, using parsimonious reasoning, and given that all other *Deinosuchus* remains are found in marine-related deposits, it is most likely that a marine lithofacies was the sedimentary setting for that individual.

The Species Namesake from North Carolina

As the chapter opening indicated, taxonomic rules can be a pain, and the case of *Deinosuchus rugosus* provides a worthy example. The holotype specimens (i.e., the individual materials on the basis of which the species is diagnosed) are two teeth, which are almost certainly from separate animals and are not associated with any other body parts. These were originally designated as "*Polyptychodon*" *rugosus* Emmons, 1858. It is generally poor paleontological practice to erect a new species solely on the basis of teeth—or at least on nonmammalian teeth. Unlike mammals, which have complex and highly derived cheek teeth, the teeth of most other tetrapods (i.e., four-legged vertebrates) tend to be simple and single-rooted, and their general makeup reflects common ancestral stems rather than the unique characters (termed "autapomorphies") that modern taxonomists prefer to use in higher-level classification. Nevertheless, where a tooth taxon has been described, and where that taxon proves to be valid, the rules of the International Committee on Zoological Nomenclature (ICZN; Ride et al. 1985) require that it be the namesake. In the case of the "*Polyptychodon*" *rugosus* teeth, we may at least be comfortable that they were correctly assigned to a new crocodylian species, because from the same deposit come unmistakable *Deinosuchus* osteoderms, large crocodylian vertebrae, and many additional teeth that we now know come from various jaw positions in *Deinosuchus*. It would still have been more desirable to have the species name-bearer consist of a more comprehensive set of fossils, but rules are rules.

The age and stratigraphic origin of the type teeth are also troublesome. The locality reported by Ebenezer Emmons (1858), the state geologist, was along the Cape Fear River, in Bladen County, North Carolina. The site was probably the locality near Elizabethtown called Phoebus Landing (Fig. 1.4), which has yielded innumerable Late Cretaceous fossils. The sedimentary origin reported by Emmons was a "Miocene marl bed" (in this sense, "marl" means a glauconitic sandstone), but Emmons recognized that these specimens were probably: "washed out of the green sand [the proximal Upper Cretaceous beds] into the (M)iocene." Subsequent descriptions and studies of what are likely the same fossil beds (e.g., Baird and Horner 1979; Miller 1967) attribute these specimens—and hundreds of other *Deinosuchus* teeth and bones discovered later—to the Black Creek Formation. This is a middle Campanian unit, which has produced many typical Late Cretaceous vertebrate fossils, including bony fish, selachians, marine turtles, dinosaurs, mosasaurs, and crocodylians. However, it is not definitively proven whether or not the holotype "*P.*" *rugosus* teeth were reworked from the Cretaceous into a Miocene deposit because they were not directly associated with age-diagnostic invertebrate fossils. And as explained below, mixed-age assemblages are to be expected in some Coastal Plain situations.

Water-worn teeth and bones, nearly always isolated from other bones, are common fossils in Atlantic and eastern Gulf Coastal Plain sandy deposits, ranging from Cretaceous to Pleistocene age. Some of these coastal deposits have fossil faunas with this range of ages all intermixed, possibly including Cretaceous, Pleistocene, and intervening age fossils. To understand the cause of this age mixing, it must first be understood that fossil bone and teeth in Atlantic and Gulf coastal sediments tend to be preserved by the calcium phosphate mineral apatite, which has a relatively high specific gravity of 3.2. The phosphorous source for the apatite is debatable, but it seems to originate from upwelling deep marine currents, which may possibly be recycling older dissolved marine bones or invertebrate exoskeleton material (chitin). The high specific gravity of apatite may be compared with that of 2.65 for the quartz sand that makes up the bulk of the sedimentary coastal deposits. The result of this density difference is that apatite-permineralized bone tends to collect as loose debris (termed "lag concentrates") in any depressions on the sea bottom, or in stream channels, oyster reefs, and other catch basins formed at various times in the past. This process also occurs in the present. One of the best methods to collect vertebrate fossils from many eastern Coastal Plain deposits is by seining streams that drain through the original deposit, focusing on holes and crevices in the stream bottom. (Unfortunately, this collecting technique does not reveal the age of the fossils, which frequently leads to uncertainties about their origins.) Because coastal deposits have been subject to frequent episodes of erosion, it follows that there have been many opportunities for bones to be eroded and redeposited in marine shore deposits.

Another set of geological processes that tend to concentrate bone and teeth are part of the transgressive phases of eustatic marine cycles, which have been a major focus of modern sedimentology (e.g., Haq et al. 1987). During marine transgressions, when seawaters encroach the land, sandy sediments tend to be poorly available for distribution on the marine continental shelves, causing, in effect, "sediment starvation." The result of sediment starvation is that phosphatic fossils become less diluted by other sedimentary materials and are especially prominent among whatever debris is deposited, especially in depressions. The results are termed "transgressive lag-type bone accumulations," and they may contain enormously high concentrations of worn, mixed bones and teeth. These deposits are frequently the targets for vertebrate fossil collecting because some of these are actual bone beds, made up of more than 50% vertebrate fossil material. Because of all the reworking that may occur during these marine erosion and redeposition cycles, Pleistocene sand deposits in several areas of North and South Carolina are known to contain a seemingly inexplicable mixture of Cretaceous dinosaur teeth, Ice Age horse teeth, and Tertiary white shark teeth. However, when we see the heavy water wear on the specimens in such deposits, their origin becomes evident.

The article by Ebenezer Emmons (1858) with the "*Polyptychodon*" *rugosus* description was a report of the North Carolina Geological Survey containing brief descriptions of a wide variety of fish and (presumably) reptile teeth in the so-called marl deposits of North Carolina. Some of the "reptile" teeth later were found to be fish teeth (Schwimmer et al. 1997a), and many of the real reptilian teeth have been reassigned from Emmon's original descriptions. Nevertheless, Emmons correctly recognized that the "*P.*" *rugosus* specimens were crocodylian teeth, and by assigning them to the genus *Polyptychodon*, he was employing a name widely used for crocodylian teeth from the Late Cretaceous English Chalk. This generic name itself has a long and complex history, but because it is not terribly important for the topic at hand, it is sufficient to state that it was discovered after Emmons' work that "*Polyptychodon*" had to be properly assigned to teeth from a pliosaur (i.e., a short-necked member of the Plesiosauridae) rather than a crocodylian. Therefore, the "*Polyptychodon*" *rugosus* teeth were generically (but not specifically, because they were truly novel) reassignable by ICZN rules.

The species name "*rugosus*" was based on the thickness of the tooth enamel and dentine, which is among the most diagnostic autapomorphies of *Deinosuchus rugosus* (Fig. 2.12). This extremely thick tooth composition is combined with many vertical striations on tooth crowns (Fig. 2.13), which tends to further increase the amount of enamel present in the tooth. The cumulative result is unusually strong teeth, especially for a crocodylian. Although Emmons could not know it when he described the teeth, *Deinosuchus rugosus* possessed an unusually wide variety of teeth in its jaws. The type teeth appear to come from the mid- and posterior jaw regions because they are large in diameter, but the crowns are relatively low. One of the type specimens would be very rounded at the tip, if it were complete (see Fig. 2.13 for

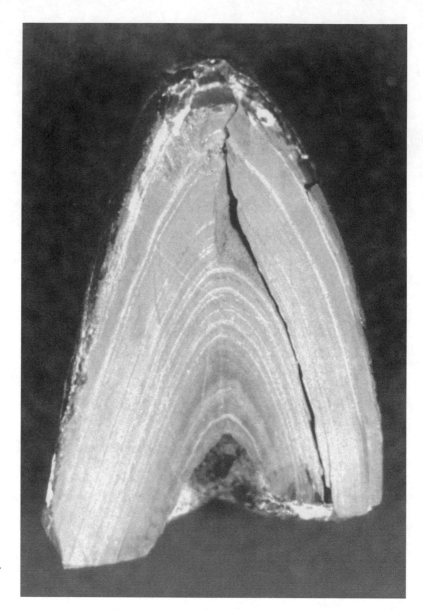

Figure 2.12. Deinosuchus *tooth from Georgia, cut to show the very heavy tissue in cross section. The width across the base is approximately 3.0 cm.*

a similar tooth). The ruggedness of the teeth, combined with the low-crowned, rounded shapes of some, suggest that the crocodylians were well adapted to feed on hard or tough prey. As will be shown in Chapter 8, this feeding behavior is in fact documented in the fossil record and is one of the most distinctive aspects of the paleoecology of *Deinosuchus*. In addition to the thickness and striated enamel, the teeth are big. The larger of Emmon's types, obviously a posterior tooth, has a large portion of the root attached and measures 8.5 cm in length and 2.9 cm across the base. The width makes it among the larger of *Deinosuchus* posterior teeth known from the eastern United States.

Figure 2.13. External views of Deinosuchus *teeth, both anterior high-crowned specimens (above) and midjaw and posterior low-crowned specimens (below). All specimens from Georgia and eastern Alabama. Note that they are all relatively blunt and feature rugose surface textures.*

As frequently happens with older taxonomic names, the teeth from North Carolina were bounced from one fossil generic assignment to another. In 1871, Edward D. Cope reviewed fossil vertebrates from North Carolina and transferred Emmons' species into an older generic name "*Thecachampsa.*" This is yet another taxonomic conundrum because *Thecachampsa* is a junior synonym of *Crocodylus* and probably cannot be valid for any Cretaceous crocodylian. Nevertheless, "*Thecachampsa*" legally replaced "*Polyptychodon*" as the generic assignment for the *rugosus* species for many years, and at least brought the teeth correctly into the realm of the more highly derived eusuchian crocodylians. Just two years earlier, Cope (1869) had described another crocodylian tooth, found originally by W. C. Kerr (the North Carolina state geologist during the Civil War), from a nearby Black Creek Formation site in Sampson County, North Carolina. This is one county over from Emmon's original "*P.*" *rugosus* locality in Bladen County. Cope (1869) dubbed the new tooth *Polydectes biturgidus,* and it turns out to be another *D. rugosus* specimen, which fortunately does not have publication priority and need not be a further problem when figuring out the proper name for the animal.

Bringing the taxonomy close to present status, three more articles addressed the assignment of these North Carolina specimens. O. P. Hay (1902), in a massive catalog covering the taxonomy of nearly all vertebrates described from North America to that date, placed the "*P.*" *rugosus* specimens into the modern crocodile genus; thus, they were redesignated *Crocodylus rugosus.* Halsey Miller (1967) reevaluated the (by then) substantial accumulation of referable giant crocodylian remains from Phoebus Landing, North Carolina, and put "*Thecachampsa rugosa*" and "*Polydectes biturgidus*" into synonymy with "*Crocodylus*" *rugosus.* The material he examined included the various type teeth mentioned, plus two vertebrae and two osteoderms from the Black Creek Formation. But he did not consider the possible relationship between these North Carolina giant crocodylians and the *Deinosuchus* and "*Phobosuchus*" specimens that had been found in the western United States.

The last and current reassignment of the giant crocodylian fossils was a minor part of an article addressing dinosaurs! Donald Baird and John R. Horner (1979) reevaluated the Cretaceous dinosaurs described from North Carolina, and among other reassignments, they recognized that a *Deinosuchus* right maxillary fragment had mistakenly been considered a theropod dinosaur fossil by Miller (1967). In fact, Miller used this alleged *Gorgosaurus* theropod specimen as one of his pieces of evidence that the Black Creek Formation shared many faunal aspects with Campanian Age deposits in Alberta and Montana (see Appendix A for detailed age terminology). Despite his misidentification of the *Deinosuchus* maxilla, Miller was correct that there are many common features between the Albertan and Black Creek faunas (see Schwimmer et al. 1993), but not regarding the crocodylians. Baird and Horner (1979) also recognized that Nopcsa (1924) had incorrectly assigned several crocodylian taxa to his genus *Phobosuchus,* including a giant Tertiary alligatoroid *Purussaurus* (discussed in Chapters 3 and 7). Therefore, "*Phobosuchus*" was invalid, and because it had publication precedence, the name *Deinosuchus* was resurrected. Thus, in 1979, Baird and Horner reassigned Emmon's "*Polyptychodon*" *rugosus* to *Deinosuchus rugosus* and restored Holland's *Deinosuchus hatcheri* back to its original generic name (because Colbert and Bird 1954 had reassigned the species to "*Phobosuchus*"). But although they discussed the possible relationships between *D. rugosus* and (the newly renamed) *Deinosuchus riograndensis* material (including that which was described by Colbert and Bird in 1954 and that which was currently under study from Big Bend, Texas), Baird and Horner (1979) did not propose that these might all represent a single species ranging from New Jersey to Montana and southwest Texas. After 1979, most knowledgeable paleontologists followed Baird and Horner in recognizing the Texas fossils to be of species of *Deinosuchus*, but most also remained unsure whether or not *D. riograndensis* is a junior synonym of *D. rugosus*. Many authors still use the generic name "*Phobosuchus*," even though it has long been proven to be invalid. Perhaps its translation of "fearful crocodile" is just too compelling to give up; but, alas, we must do so.

3. The Size of *Deinosuchus*

Estimating the Size of *Deinosuchus* and Other Crocodylians

The story of a "giant" crocodylian invariably brings up the questions "How big was it?" and "How does it compare with other giant animals?" Neither of these questions may be answered without some initial considerations because the concept of "giant" organisms is not a simple one. Consider, for example, the relative sizes of a common tree, such as a white oak, compared with the largest living land carnivores, such as tigers and brown bears. The tigers and bears seem huge and fearsome to 75-kg humans, yet they are orders of magnitude smaller than an ordinary, unimposing tree. Some of the longest organisms on Earth are kelp, and apparently the largest living organisms both in dimension and mass are species of soil fungi. An insect or spider that approaches a kilogram in weight seems gigantic and fearsome to most of us because it is so far from the expected. (Consider the sensory impact of a big tarantula dropping on your head, or the geeky fascination some people have with "giant" Asian hissing cockroaches.) Similarly, we are awed and terrified by the size of a mounted skeleton of *Tyrannosaurus* and other giant theropods (including several newly discovered taxa of equal and larger size, such as *Giganotosaurus, Carcharodontosaurus,* and *Suchomimus*)—yet they were all smaller than the largest individuals of *Orcinus orca,* a modern killer whale. Ironically, orcas are fierce predators that small children are sometimes invited to pet in water parks.

Crocodiles and alligators, in general, are among the larger living predatory animals. Some crocodile species are the largest living terrestrial vertebrate predators to survive since the Pleistocene Ice Age (and see discussion to follow in this chapter about other fossil giants). The largest sizes reached by modern crocodiles can only be judged from recent historical reports, circa the 1800s to the early 20th century, because it is generally recognized that recent human competition for habitats and the trade in wild crocodylian skins have limited the growth potential of all species. For better or worse, relatively small humans are better predators than are even the largest crocodylians; and, unfortunately for our purposes, historical reports of animal size are not reliable unless they are documented with actual measurement data or preserved materials (usually skulls). In addition, most reports of great size in crocodylians were based on total length measurements, with few weights recorded. This is quite understandable, given that the largest modern crocodiles and alligators may reach a ton in weight and are rarely cooperative in carrying themselves to a convenient set of scales. It is obvious that length measurement is much simpler because a total length measurement can be taken with a simple measuring tape or by comparison with an object of known length (for example, a boat).

Eugene Meyer (1984) summarized records of large crocodylian measurements and reported that *Crocodylus* species have been reliably measured at total lengths of 6.1 m for an Indo-Pacific *Crocodylus porosus* (the saltwater crocodile) and 5.5 m for *Crocodylus niloticus* (the African or Nile crocodile); an early 19th-century measurement of an Orinoco crocodile (*C. intermedius*), a relatively slender-snouted species, was reportedly 22.3 feet (6.7 m). Hugh Cott (1961) reported 20th-century Nile crocodile measurements from southern Africa at between 18- and 22-foot lengths (5.8 to 6.5 m), many of which were made by commercial skin hunters, whom we presume measured the skins accurately for sale. Cott stated that 20 feet (6.1 m) was an indisputable length reached by the species, and several measurements to 22 feet (6.7 m) seemed reliable. The American alligator, *Alligator missippiensis,* is a shorter-skulled crocodylian and therefore will have lesser total lengths for animals that are really quite massive. The largest measured modern alligator specimen reported by Meyer (1984) was 5.84 m in length and would probably have a body mass equivalent to a 6.0-m *Crocodylus* species. However, Woodward et al. (1995) noted that no materials were preserved from this and other unusually large reported Louisiana alligators. They cited what they considered reliable reports of Florida alligators that were measured at 5.0 and 5.3 m.

From these data, one observes a trend among modern crocodylians from many species to reach a maximum size topping out around 6.5 m in length. Although Nile and Indo-Pacific crocodiles have been reported to reach lengths of 8.0 m in stories and second-hand reports, there is no credible reason to assume these reports are true. Greer (1974) showed that a well-known 19th-century report of a 33-foot (10.0 m) *Crocodylus porosus* specimen, whose skull reposes in the British Museum (Natural History), in fact turns out to be under 6.0 m length when one extrapolates the whole animal size from the skull by use of scaling curves as discussed below. The skull of a very large, living crocodylian appears long and massive to anyone's eyes, and even a 5.0-m animal will have a skull longer than a full-size, large carnivorous dinosaur such as an *Allosaurus* or an *Albertosaurus*. Crocodylian skulls are also thick and bulky, presenting a massive appearance. It is not surprising that even professional scientists tend to exaggerate the overall size of crocodylians on the basis of observations of only the skull or mandibles.

Because no complete, adult *Deinosuchus* individual has yet been described, we are forced to make extrapolations of their possible total lengths from the parts we know (Schwimmer 1999). Among the fossils that will allow useful lengths measurement are several complete jaws, which will have been slightly longer than the skulls they accompanied. A nice feature of crocodylian anatomy is that we can closely estimate the skull length from its accompanying mandible; this is done by measuring the sagittal (i.e., midline) distance from the front of the mandible, at the symphysis, to the midpoint of a line perpendicular to the center of the articular surface (see Fig. 3.1 for anatomy of the crocodylian head). This lower jaw measurement is equivalent to measuring the skull from the tip of the snout (the end of the premaxillae) back to the supraoccipital bone (immediately behind the parietal bone). This posterior point is the back of the skull table and has been used as a standard anatomical reference by most studies of crocodylian skull length. The actual rearmost points of the skull are the ends of the quadrates, just as the retroarticular processes are the rear points of the mandibles, but these structures themselves vary among crocodylians, and so the standard measurement points have evolved because they are relatively uniform. We also have several *Deinosuchus* skull specimens where both quadrates are intact and preserved with their correct spacing, allowing accurate measurement of the posterior skull widths. It is also possible to extrapolate crocodylian body lengths from vertebrae and limb bones, but these are less certain parameters and lack the prior study and literature support that we have for skull-to-total length estimates. Nevertheless, in the discussion that follows, I will on occasion refer to lengths of vertebrae as an argument for estimating the sizes of *Deinosuchus*.

The "*Phobosuchus*" *riograndensis* type material (Chapter 2) was very incomplete, and the Colbert–Bird skull and mandible reconstruction, as discussed before, was too wide, too strongly tapered, probably too high (laterally), and definitely too big overall. Colbert and Bird (1954) estimated the length of the specimen at 50 feet (15.2 m), which has been frequently repeated as the size of *Deinosuchus* in both popular books and more sophisticated paleontological literature (e.g., Carroll 1988; Meyer 1984; Romer 1966). The reconstructed total skull length, and thus the body length of the specimen, is especially difficult to judge because much of a crocodylian's skull length comes from the rostrum (the area anterior to the eyes), which in turn depends in large part on the length of the tooth row of the maxilla. From new fossils of *D. rugosus*, we now know that the species had a relatively long maxillary tooth row with 21 or 22 teeth, whereas the Colbert–Bird reconstruction had only 13 teeth in the maxilla (Fig. 2.5). This might suggest that the reconstruction was actually too short, except that for some reason, their model was also created with unnaturally wide spacing between the teeth and a very long postmaxillary and postdentary area (i.e., the regions posterior to the teeth). From newer specimens, we know that

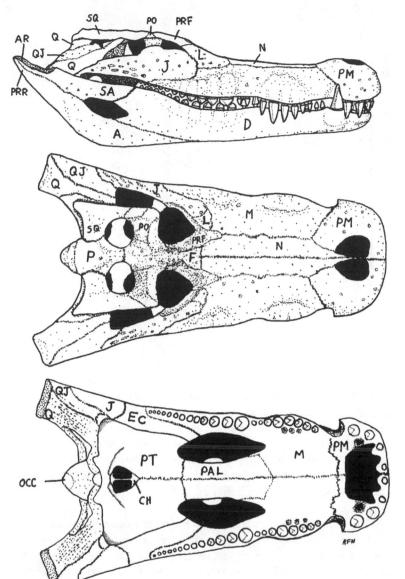

Figure 3.1. Bones of the reconstructed Deinosuchus *skull. (Top to bottom) Lateral, dorsal, ventral views.*

Skull, dorsal:
AN = *angular;*
EC = *ectopterygoid;*
F = *frontal;*
J = *jugal;*
L = *lacrimal;*
M = *maxilla;*
N = *nasal;*
P = *parietal;*
PM = *premaxilla;*
PO = *postorbital;*
PRF = *prefrontal;*
PRR = *retroarticular process;*
Q = *quadrate;*
QJ = *quadratojugal;*
SQ = *squamosal.*

Skull, ventral: CH = *choanae;*
OCC = *occipital;*
PAL = *palatine;*
PT = *pterygoid.*

Mandible: A = angular;
D = *dentary;*
SA = *surangular.*

Drawings by Ron Hirzel, modified from Iordansky (1973).

the teeth in *Deinosuchus* are quite closely spaced and that the teeth taper in size posteriorly and reach far back toward the posterior. In fact, the small, rearmost upper teeth are right up against the suture between the maxilla and ectopterygoid bone (Fig. 3.1).

In truth, it is nearly impossible to come up with a reliable skull reconstruction from the "*Phobosuchus*" *riograndensis* type specimen by itself, simply because the remains are too scanty. (However, it is still an important set of fossils for size estimation, as they are among the larger deinosuchids.) By comparing the proportions of this individual's premaxilla and vertebrae with more complete material to be considered, we can suggest how much larger the incomplete large animal was

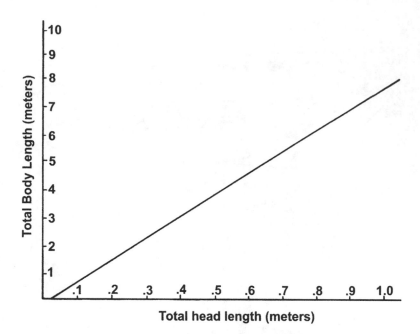

Figure 3.2. Regression line comparing total head length and total body length in Crocodylus porosus *(after Greer 1974). Redrawn from Greer (1974); data converted to metric system.*

than the more complete, smaller specimens. It is unfortunate that the Colbert–Bird reconstruction has been the standard on which so many authors have based estimates of the length and (by further extrapolation) the weight of *Deinosuchus*. The single "*Phobosuchus*" *riograndensis* dorsal vertebra associated with the type specimen is about the same size as the dorsal vertebrae from the Montana *Deinosuchus hatcheri* type specimen, suggesting that these two animals were of comparable size. No other parts of the *D. hatcheri* specimen are useful for size extrapolation, although they come from an obviously huge crocodylian.

There has been a tendency in some recent considerations of *Deinosuchus*'s size to do the reverse of older concepts and minimize the largest total lengths and weights. Ross (1989) cited an unnamed source estimating the overall length at 11.0 m. He based this estimate on a "1.8-m jaw from Texas" which may be the "*Phobosuchus*" *riograndensis* type reconstruction, or it may be a second specimen from Big Bend, held in the University of Texas collections. As will be discussed below, if a *Deinosuchus* jaw were reliably known to be 1.8 m long (a fact not yet established), then the total length of the animal would be larger than 11.0 m. Ross also estimated the maximum weight of that same animal at 6 tons. Most recently, Gregory Erickson and Christopher Brochu (1999) discussed growth in *Deinosuchus riograndensis* (see the discussion at the end of this chapter) and suggested that the maximum size the species reached was around 10.0 m. To put in perspective the difference between a 10.0- and a 15.0-m crocodylian, at the extremes of range estimates for *Deinosuchus,* consider that the weight of an individual at the smaller size estimate would be around 4.0 tons, whereas the larger would be well over 15 tons!

Fortunately, there are published data for living crocodylians that allow reasonable estimations for extinct species, and not surprisingly, the outcomes of such estimates range between the larger and smaller of the sizes which have been cited in the past for *Deinosuchus*. From total head length (THL) measurement of a crocodylian, one may extrapolate the total body length (TBL) in instances where the growth curves of the species are known. They obviously are not known for extinct *Deinosuchus rugosus*; however, numerous studies have been done on several living species of *Crocodylus* and *Alligator mississippiensis,* and we may work with these to approximate the TBL of *Deinosuchus.* Greer (1974) plotted such data for *Crocodylus porosus* (Fig. 3.2), reporting measurements from a range of specimens of less than 2 feet (0.6 m) up to 16 feet (5.2 m) in TBL. He also included THLs of four additional very large skulls from museums, without their accompanying bodies. His purpose was to test whether these latter animals would have been as large as their legends claimed (up to 10 m). His analysis of many specimens produced a plot that allowed a well-fitting regression line, showing that the relationship was nearly linear, with the formula (using the British System measurements in inches):

$$TBL = -4.39 + 7.49 THL$$

By this formula, the original "*Phobosuchus riograndensis*" skull reconstruction with a 2.0-m jaw length extrapolates to a crocodile with TBL of 41.0 feet, or 12.5 m (Fig. 3.2). Of course, the accuracy of this extrapolation depends on both the accuracy of the skull and jaw-length reconstruction and the assumption that the skull-to-body length relationship in *Deinosuchus* matches that of *Crocodylus porosus.*

Woodward et al. (1995) calculated a method for extrapolating TBLs from THLs of *Alligator mississippiensis.* Their scaling curves were based on a large sample of Florida alligators for which they calculated separate male and female parameters, because it is well known that there are many size differences due to sexual dimorphism, sufficient to cause modest errors in overall size estimates. Their scaling curves are slightly more refined than those used in Greer's method, because they used natural logarithms for a best-fit approximation of the data. For male *Alligator,* Woodward et al. stated the relationship as follows (all measurements in centimeters):

$$\log_e(TBL) = 2.132494 + 0.95811(\log_e(THL))$$

By this formula, the Colbert and Bird (1954) *Phobosuchus* skull reconstruction scales to a TBL of 12.2 m. This shorter length estimate compared with Greer's method may result from the fact that alligator skulls are relatively short compared with species of *Crocodylus,* and thus the total length will be shorter. New specimens of *Deinosuchus* from the Big Bend region are currently in preparation and study at the University of Texas Laboratory in Austin, under the direction of Professor Wann Langston Jr. (see previous discussion in Chapter 2). At least one of these is well preserved and in sufficient state of preparation to produce a measurable THL, which Professor Langston has reported as ~1.31 m (W. Langston Jr., personal communication). Plugging this into Greer's formula, this produces a TBL of 9.8 m. Although this specimen is not necessarily at the largest size of the species, it gives another benchmark estimate. It may also be the specimen on which Erickson and Brochu (1999) made their estimate of 10.0 m as the maximum size of *Deinosuchus,* because this is the longest skull known that preserves the snout-to-supraoccipital distance.

Two *Deinosuchus* specimens in my research collections from the eastern Gulf of Mexico region (one each from Alabama and Georgia) have mandibles that are nearly complete (Schwimmer and Williams 1993) from front to back (Fig. 4.7), therefore allowing very reliable measurements of THLs. These two eastern mandibles are of nearly identical size, and they coincide in width and other proportions with two additional partial skulls, one from Mississippi and one from Georgia. These are all of such similar sizes (within 5% in all proportions) that they would seem to represent the typical mature sizes of eastern Gulf *Deinosuchus rugosus* populations. There are also literally dozens of jaw fragments known from Georgia, Alabama, and North Carolina with the same proportions, along with hundreds of teeth that would fit into those size of jaws, reinforcing the idea that these are the typical regional sizes (i.e., in the southeast). It is also clear these are not among the larger *Deinosuchus* specimens known because the THLs they represent are slightly over 1.0 m (~39.0 inches). However, it is worthwhile to calculate their TBLs because these will represent the average size at death for eastern Gulf Coast deinosuchids. By use of Greer's formula, these mandibles correspond with TBLs of ~8.0 m (26.2 feet), which apparently was the average mature length of the eastern *Deinosuchus* population. Later in this book, in Chapters 6 and 8, I will speculate why the eastern deinosuchid population seems to consist of mostly smaller animals. As a hint, I propose that it is related to the geologic age (older) in the east and to their differing feeding habits. An interesting, but probably remote, possibility also exists that the eastern population consisted of females, whereas the western population comprises mixed-sex populations or mostly males, which can be assessed on the basis of the well-known sexually dimorphic size differences in alligator populations. This scenario too will be discussed later.

There are other measurements, beside head lengths, that suggest that western *Deinosuchus* reached much larger sizes than the 8.0 m typical of the eastern population. For example, *Deinosuchus* dorsal vertebrae I have observed from many eastern localities are a maximum of 97.0 mm long sagittally (lengthwise from ball to socket). To my knowledge, no specimen has been found along the Eastern Coastal Plains larger than this. Now, if one accepts the assumptions and calculations showing that the average TBLs of eastern *Deinosuchus rugosus* are 8.0 m and that this TBL corresponds with vertebrae of a maximum of 97 mm in length, we may compare these measurements with vertebrae from the same anatomical positions in the *Deinosuchus hatcheri* type specimen, measured at 150 mm in the same dimensions (Holland 1909). Also, a similarly sized vertebra accompanied the "*Phobosuchus*" *riograndensis* skull material, and even larger vertebrae have been collected in the Big Bend region by Professor Langston's cowork-

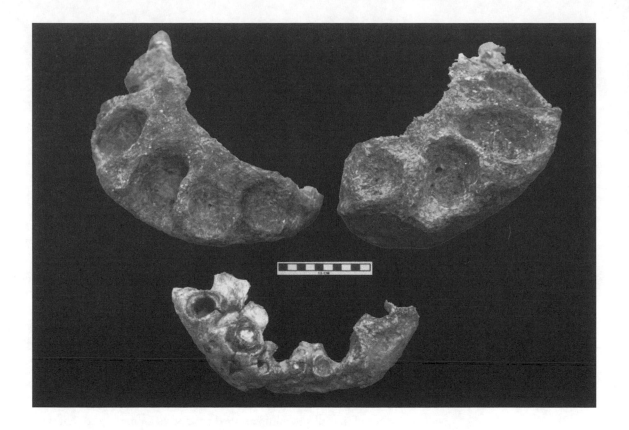

ers (W. Langston Jr., personal communication). Comparisons between vertebral dimensions and TBL have not been graphed and published for modern crocodylians, as have the skull lengths. However, from a modest sample of modern *Alligator* specimens ranging from 1.0 to 3.2 m, I have observed that vertebral lengths track closely with overall length. From this, I deduce that vertebrae 1.5 times as big as the typical eastern *Deinosuchus* vertebrae correspond with crocodylians (1.5 × 8.0 m = 12.0 m) in TBLs.

This estimate of 12.0 m also corresponds with my observations on the general proportions of comparable bones of eastern and Big Bend *Deinosuchus* skulls. For example, Figure 3.3 shows that the premaxilla of a well-preserved 8.0-m specimen from Alabama is dramatically smaller than the American Museum of Natural History (AMNH) "*Phobosuchus*" holotype premaxilla. It is true that the AMNH specimen is damaged in the midline and therefore its actual width is uncertain: nevertheless, the photograph shows how much bigger it really is: nearly twice as large in all proportions. Even if there were a significant amount of disproportional size change in various body parts among the very different sized animals (a phenomenon technically termed "allometry"), I am convinced that the premaxillae, as well as all other bones I have examined, show that the AMNH specimen was very much larger than the 8.0-m Alabama specimen. The *Deinosuchus hatcheri* type specimen also could not have been smaller than the AMNH type, given

Figure 3.3. Comparison of two Deinosuchus *premaxillae to show the extraordinary size differences between representative Big Bend and eastern specimens. (Top) "Phobosuchus" riograndensis holotype (in two pieces) from Big Bend. (Bottom) Alabama specimen from the Mooreville Formation (Jones Dam Site). Scale in centimeters. Original photographs by D.R.S., with considerable computer enhancement by Jon Haney.*

the size of the vertebrae. Until a full-length skull specimen from Texas or Montana is found and reported that equals the largest known *Deinosuchus* materials (which are still the types of "*P.*" *riograndensis* and *D. hatcheri*), I presume that vertebral length comparisons give us the closest approximations we can use for extrapolating maximum lengths. I therefore predict that a 12.0-m maximum TBL for *Deinosuchus* will turn out to be very close to correct, if we are ever fortunate enough to discover a complete specimen from the largest size range.

We can go farther in size estimation and extrapolate body weights from our estimated TBL calculations. Here the numbers become quite amazing, because crocodylians are well known to increase their mass exponentially (and quite strikingly) with increasing length. As explained, weights are rarely recorded for big crocodylians because of the difficulty of doing so. Among the greater crocodylian weights reported in the literature, Cott (1961) stated that the weight of a *Crocodylus niloticus* specimen of 21.5 feet (7.0 m) was approximately one British ton (~900 kg). Going to larger-sized individuals by extrapolation from measurements of captive populations of *Crocodylus porosus,* Webb and Manolis (1989) predicted weights for specimens ranging up to 10 m. Thomas Coulson et al. (1973) did similar studies with small, captive Louisiana alligators and produced a total length to body weight curve with the following formula (in centimeters and kilograms):

$$\log_{10} \text{(body weight)} = 3.35 \log_{10}(\text{TBL}) -6.10$$

They noted that this formula applied to captive animals, and that wild populations would likely be leaner and hence weigh less. They also stated that the curve might not track correctly for animals over 3.0 m in TBL because the largest animal in their direct study was less than 2.0 m long.

By use of Webb and Manolis's tabled estimates, the typical 8.0 m *Deinosuchus* of the eastern Gulf Coast would weigh in at about 2300 kg, or 2.3 tons (about the size of a mature *Daspletosaurus* theropod; Fig. 3.4). The Coulson et al. formula yields weight estimates for these smaller *Deinosuchus* of ~4.2 tons each. The Big Bend "*Phobosuchus*" *riograndensis* type specimen is well beyond Webb and Manolis's estimations, but by further extrapolating their extrapolations, I calculate that a 12.0-m *Deinosuchus* had a body weight of 8.5 tons! This was at least 1.5 tons larger than the typical published estimates for the weight of *Tyrannosaurus rex,* which has traditionally been the giant terrestrial carnivore benchmark. The formula published by Coulson et al. comes out to an astounding tonnage of 16.4 metric tons for a 12.0 m TBL, which is probably inapplicable because of the cautionary notes they mentioned (especially extrapolating from 2-m to 12-m animals!). Nevertheless, it does suggest that the 8.5-ton estimate is conservative.

Summarizing these calculations, the largest known *Deinosuchus* specimens, both the Big Bend "*Phobosuchus*" *riograndensis* and the Montana *D. hatcheri* holotypes, would have been about 12.0 m long and weighed at least 8.5 tons. The common *Deinosuchus* specimens found along the eastern continent averaged much smaller sizes but still were typically 8.0 m long and weighed about 2.3 tons. All of these

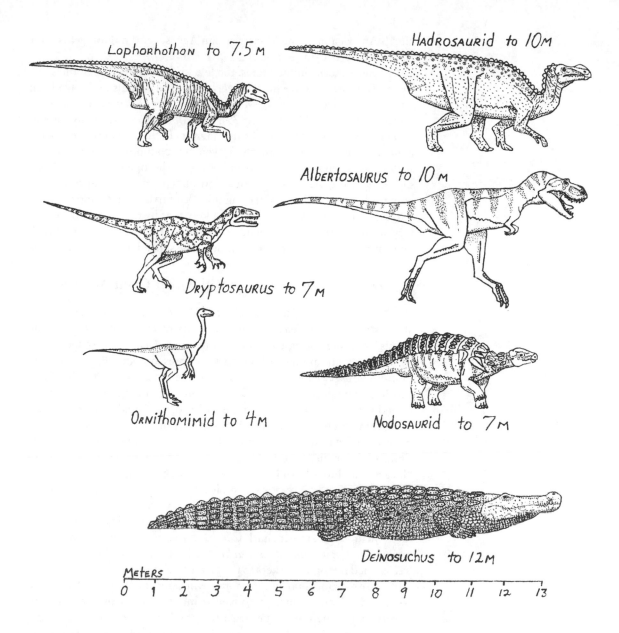

Lophorhothon to 7.5 m

Hadrosaurid to 10 m

Albertosaurus to 10 m

Dryptosaurus to 7 m

Ornithomimid to 4 m

Nodosaurid to 7 m

Deinosuchus to 12 m

Meters

| 0 | 1 | 2 | 3 | 4 | 5 | 6 | 7 | 8 | 9 | 10 | 11 | 12 | 13 |

crocodylians were sufficiently large to kill and eat any contemporary Late Cretaceous nearshore and marine life, with the possible exception of the larger mosasaurs, plesiosaurs, and pliosaurs. Deinosuchids on both sides of the continent were at least equal in size—and usually larger—than any known land predators of their age, including the predatory dinosaurs. The theropods that were contemporaries of *Deinosuchus* in the west (and also found in the same nearshore marine strata in Texas, Wyoming, and Montana) were small dromaeosaurids and medium-large tyrannosaurids up to the size of *Daspletosaurus* and *Albertosaurus* (Horner 1989; Lehman 1997; Rowe et al. 1992). (The largest North American theropod, *Tyrannosaurus,* did not appear in

Figure 3.4. Comparison of the size of largest credible western Deinosuchus *with other representative Late Cretaceous tetrapods. The theropod* Albertosaurus *is close to the (so far) undescribed large theropod species found in association with* Deinosuchus *in both the Aguja Formation and the southeastern United States. Drawing by Ron Hirzel.*

North America until the Maastrichtian Age, several million years after *Deinosuchus* became extinct.)

In the eastern United States, the few known theropods of the same age as *Deinosuchus* are all relatively small individuals and have been referred to *Gorgosaurus* by Miller (1967), *Dryptosaurus* or *Albertosaurus* by Baird and Horner (1979), and *Albertosaurus?* by myself and others (1993). Regardless of the true identity of these eastern theropods, none was more formidable than even the common, smaller *Deinosuchus* with which they coexisted. And as the opening of this book suggested, the eastern carnivorous dinosaurs themselves may have been prey to the crocodylians. Carrying this idea farther (and a more extensive discussion is in Chapter 8), it is likely that by their superior predatory abilities, the giant crocodylians were affecting the carnivorous dinosaurs' ability to compete for food in the shore-based environments.

Deinosuchus Compared with Other Giant Animals

Now that the length and weight of *Deinosuchus* have been reasonably established, at least within the limits of extrapolation from the fossils, it is interesting to compare the larger *Deinosuchus* specimens with other giant animals of the past and present. Among the most astounding observations we may make at the outset is that the Big Bend and Montana populations of *Deinosuchus* are among the largest four-legged terrestrial carnivores that ever lived. The qualification "seem" in that statement is not based on doubts about their size, but rather on the possibility that *Deinosuchus* was not a truly terrestrial animal (see Chapter 5 for discussion of its habitats). In this discussion, the term "largest" includes both length dimensions and weight.

Giant animals are always subjects of interest (see, for example, Paul 1997) and are often subjects of misjudgment about their very size: for example, two notorious examples of miscalculated giant-sized creatures include the extinct shark *Carcharodon megalodon* and the Pleistocene ape *Gigantopithecus blacki*. Teeth of *Carcharodon megalodon* (reassigned by some workers to the genus *Carcharocles*—e.g., Cappetta 1987) are extraordinarily common in Miocene and Pliocene deposits in North and Central America, as well as many other areas with contemporary marine deposits. The larger *C. megalodon* teeth are 20 cm from tip to root, and when compared with the teeth of living great white sharks, would suggest individuals with body lengths reaching 20 m. There have been many jaw reconstructions made from the teeth of *C. megalodon,* including the famous AMNH display that was photographed with several personnel comfortably (and smugly) enclosed by the jaws. Unfortunately, this amazing reconstruction represents a gross exaggeration of both the jaw size and the shark's overall size. The reconstruction assumed the tooth spacing in *C. megalodon* was similar to that of *C. carcharias* (the modern great white shark) and that there was relatively little size differentiation from midline to hinge on the jaws. Thus, the AMNH reconstruction (and many others that followed it) attempted to fit huge, widely spaced teeth around most of the jaws, rather than strongly tapering them toward the jaw hinge, forcing the

Figure 3.5. Modern reconstruction of Carcharodon megalodon *jaws, width approximately 1.2 m.*

result to be about twice the natural size. From the oversized jaw reconstruction came the inevitable oversized length and weight estimates. Recent work (Fig. 3.5) and reconsiderations (see Gottfried 1997), especially those based on preserved *Carcharodon megalodon* vertebrae, indicate that the larger individuals were about 16 m long. These were still incredibly huge and formidable sharks, but no larger than modern killer whales.

The overages in estimating sizes in the fossil ape *Gigantopithecus blacki* are less egregious, but just as interesting, because *G. blacki* was once considered a possible close relative of humans. (And thus it was believed for a time during the early 20th century that we had a gigantic ape in our family tree!) In brief, *Gigantopithecus* was first described solely on three teeth found (and purchased) in a Chinese apothecary shop by the German paleoanthropologist G. H. R. Von Koenigswald in 1935. The teeth included huge molars, some more than 3 cm in width and equally thick and high in proportion. On the basis of these teeth, Von Koenigswald and later workers, who discovered Miocene-age jaws with similar teeth, extrapolated that they came from an ape weighing 600 kg and having a height of nearly 3.0 m. The human–ancestor connection for *Gigantopithecus* was based on a once-common paleoanthropological perception that Miocene Asian apes were the linear ancestors of hominids (i.e., the family of humans). The basis of this presumption was two shared derived characteristics of early humans and Asian apes: large molars and V-shaped dental arcades (jaw out-

Figure 3.6. Mounted skeleton cast of Giganotosaurus carolinii, *the largest known predatory dinosaur. The total length of the specimen is approximately 13.0 m.*

lines). The Asian apes include the living orangutans, the fossil ramapithecines, and *Gigantopithecus*. The Asian ape–hominid association was effectively disproven when molecular studies in the late 1980s demonstrated that humans were clearly descended from African ape ancestors and when new fossils showed that *Ramapithecus* and *Gigantopithecus* were closer to modern orangutans than to the African apes. Newer fossils, including skulls and jaws from Chinese caves (rather than drugstores), also showed that *Gigantopithecus* lived through the Pliocene and into the Pleistocene. Further, and significantly for the point here, this group of Asian primates had huge jaws that were disproportionately large. The overall body sizes were not much larger than large gorillas (approximately 300 kg; Fleagle 1988), although the lower jaws were nearly twice as massive. So *Gigantopithecus* was neither bizarrely huge (for an ape) nor a hominid ancestor.

As the weight calculations show, the larger *Deinosuchus* remains extrapolate to tonnage of at least 8500 kg and lengths of 12.0 m. No living terrestrial predator comes even close to that size, and among extinct animals, only larger carnivorous dinosaurs and other crocodylians are near competitors. Newer discoveries of giant theropod dinosaurs from the early Late Cretaceous of the Southern Hemisphere, both of Cenomanian Age (see Appendix A), raise the bar just slightly above the size of *Tyrannosaurus rex,* long considered the largest theropod. *Giganotosaurus carolinii* from Argentina (Coria and Salgado

1995), according to the authors, extrapolates to a body length of 12.5 m and a weight of 6 to 8 tons. This size is extrapolated from an incomplete specimen (Fig. 3.6), but that specimen does include a nearly complete pelvis, partial legs, most vertebrae, and enough skull material to give a good basis for reconstructing the animal's proportions. Another huge theropod, *Carcharodontosaurus saharicus* from Morocco (Sereno et al. 1996), has been known from fossils for 70 years, but a recent skull recovered and described by Sereno et al. is well preserved and 1.6 m long. Given the known postcranial proportions of the animal, this skull extrapolates to an overall body size just slightly longer than the longest *Tyrannosaurus* specimens, and of about the same weight as *T. rex*. These new giant theropods still extrapolate to tonnages that are slightly less than the larger *Deinosuchus* and to TBLs that are no greater.

Crocodylians frequently grow to large sizes. This fact may be explained by several aspects of their makeup and habitats, especially their association with water. Clearly, an animal living mostly or entirely in water requires less supporting structure than does a terrestrial animal, and it may grow to sizes unsupportable on land. Larger whales are perfect examples of this relationship, as are large mosasaurs (Late Cretaceous marine lizards) and plesiosaurs (a separate, diverse group of Mesozoic marine reptiles). It is unlikely that any of these large marine vertebrates could have ever successfully walked on land, just as modern whales usually die when they are beached or caught in water too shallow to float their bodies. The relationship between aquatic habitat and large size does not mean that gigantic land animals are improbable —sauropod, theropod, hadrosaurian, and ceratopsid dinosaurs, elephants, and certain fossil rhinoceros prove that they have existed—but it does suggest that land animals must contend anatomically and evolutionarily with many biomechanical problems if they are to move freely. Animal bodies that are suspended on dry land solely by muscle and bone require more complex adaptations than do those that take advantage of the buoyancy provided by water.

But crocodylians commonly spend part of their lives on land, and many are able to move swiftly, taking advantage of several biomechanical specializations. Here too we see the possibilities for large sizes favored by several uniquely crocodylian morphologies. Chief among these are the reinforcing structures in their dorsal skins: the thick osteoderms, which were introduced in Chapter 2. As explained there, the 'derms serve primarily as weight-supporting, external, muscle-encased buttresses (Fig. 3.7), which allow the bulky bodies to be slung low over their short legs, and yet which enable them to lift off the ground surface for overland locomotion by the tensional force exerted across the animals' backs. Without these 'derms and their ligamentous and muscular associations, the high-walk and "gallop" modes of locomotion seen in even larger modern *Crocodylus* (Cott 1961; Frey 1984) and *Alligator* (Meyer 1984) would not be possible.

The thick squamation (scales) across the ventral (bottom) side of almost every crocodylian species also facilitates land motion by larger-bodied individuals. These durable, protective surface coverings allow the large animals to slide on their bellies ("the belly run," in the termi-

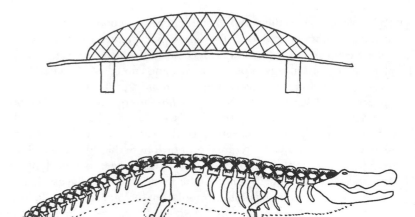

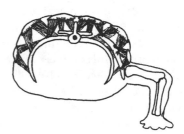

Figure 3.7. Illustration of the boxwork bridge analogy for the function of crocodylian osteoderms. The cross-bracing musculature across the dorsal surface, inserted into the 'derms, helps to elevate the abdomen during high-walk locomotion. Modified from Frey (1984); drawing by Ron Hirzel.

nology of Cott 1961), at running speed or even faster, without damaging skin or underlying organs. Hugh Cott observed modern Nile crocodiles belly-sliding rapidly down steep 40- to 50-foot (13–16 m) river banks, like so many toboggans down a snowy slope, hitting the river water with tremendous splashes, to attack prey or to escape when surprised and alarmed. This locomotion is more rapid than the usual high walk and also uses very little energy in the process. More significantly, it requires virtually no need for the use of the limbs and therefore allows another rapid motion of a large-bodied, short-limbed animal. Here we see again that basic crocodylian character enables land motion in an animal that might appear too massive to move effectively.

Given that crocodylians have both habitat preferences and evolutionary preadaptations that allow them to achieve large sizes, it is not surprising that large and gigantic taxa evolved repeatedly through the history of the group. At several times during both the Mesozoic and early Cenozoic Eras, several marine crocodylians evolved elongate forms that achieved great lengths (see Chapter 7 for more comprehensive discussion of deinosuchid ancestry and relatives). As one might predict in species largely dedicated to aquatic life, the importance of the osteoderms was usually minimized in some species, and some of the marine groups, such as the Upper Jurassic family Metriorhynchidae, completely lost their dorsal osteoderms. The metriorhynchids, which lived in the mid-Mesozoic, reached only modest sizes, around 5 m. However, another marine crocodylian family, the Dyrosauridae, includes at least one species, *Phosphatosaurus gavialoides* of the Eocene of northwest Africa (Buffetaut 1979), reported to reach 9.0 m in length. Despite the great length, *P. gavialoides* was far more slender than *Deinosuchus,* especially in the proportions of the skull that was elongate and narrow-jawed, and it certainly was not half as massive. *Phosphatosaurus gavialoides,* coincidentally, converged on another morphology of *Deinosuchus* in having robust, blunt teeth, especially in the posterior part of the mouth. As will be discussed in Chapter 8, both *P. gavialoides* and *Deinosuchus rugosus* have been inferred to have eaten marine turtles as a preferred food on the basis of the tooth morphology and, in the case of *D. rugosus,* other evidence.

At least two partly terrestrial crocodylomorph species equaled *Deinosuchus* in both length and weight. One was *Purussaurus brasilicus,* a Miocene caiman from the Amazon basin in Peru and Brazil (Campbell and Frailey 1991). This genus has been known for more than 100 years (Gervais 1876; Nopcsa 1924), but recent specimens, especially Campbell and Frailey's reconstruction of a well-preserved skull (Fig. 3.8), showed these alligatoroids to be broad-headed, high-skulled monsters with 2-m skulls, estimated TBLs of at least 12 m, and estimated weight of at least 10 tons. A jaw fragment from the same genus has been found that extrapolates to an even larger individual than the 2-m skull, with a TBL estimated at slightly over 13.5 m. The second species co-equaling *Deinosuchus rugosus* in overall size is *Sarcosuchus imperator*, recently redescribed by Sereno et al. (2001) based on new specimens from the mid-Cretaceous of Africa (Niger). *Sarcosuchus* was a longirostrine (narrow-snouted) basal relative of the crocodylians (hence my use in the paragraph opening of the more general term "crocodylomorph," discussed in Chapter 7). Sereno et al. hypothesized that its overall body morphology and habitat was generally similar to the more derived river-dwelling true crocodylians, and they estimated the length of a large individual at 11 to 12 m, with the weight at 8 tons. As with *Deinosuchus*, larger individuals of *Sarcosuchus* seem to have evolved their size because they had abundant giant prey, as considered in the discussion below. *Sarcosuchus* is further discussed in Chapter 7 in terms of its taxonomic relationships.

Why Did *Deinosuchus* Become So Large?

Scientists assume that Nature is parsimonious and therefore that the characteristics of creatures are likely to have appeared and persisted for good reasons. Of course, there may be a lot of randomness in evolution, at least according to modern perceptions of the theory, but extreme morphologies probably occur as adaptations to specific conditions. In that light, we may ask why *Deinosuchus* evolved its noteworthy characteristic of such giant size. And we may also extend the question to ask why some *Deinosuchus* populations seem to have contained many of the larger individuals (e.g., especially the Big Bend fossils), whereas some regions seemed to have supported mostly smaller indi-

Figure 3.8. Reconstruction of the skull and mandible of Purussaurus brasilicus, *a gigantic fossil Tertiary caiman from South America.*

viduals (e.g., the eastern Gulf and Atlantic Coastal Plain populations). Does the fossil record supply or suggest answers?

Organisms, especially predators, usually evolve the sizes they need for efficient feeding balanced against the needs to maintain their life functions. Reason and logic tell us that an oversize predator will have to kill a lot of prey to maintain its mass, whereas one that is too small may not be able to kill with sufficient effectiveness to provide for its needs, and it may miss out on high-grade food resources. This is overly simplistic, of course, because it is obvious that carnivores must mature and grow, and therefore many species, especially among lower animals where there is little parental care, must vary their predatory strategies as they grow. From observations on modern species, we can document the changes in prey that many predatory animals undergo as they mature: for example, young tiger sharks tend to hunt schooling fish, whereas large tiger sharks will kill virtually any animal they can handle, including seals, porpoises, and large fish. It is inevitable that *Deinosuchus* varied its predatory habits as it matured too; but yet we find entire geographical regions that seem to host one size class of the crocodylians (e.g., smaller in the eastern Gulf Coast), and surely these are not collections of all mature or immature individuals. There must have been some region-specific environmental conditions or prey availability that affected the average mature sizes of individual *Deinosuchus*.

If the overall size and killing efficiency are assumed to have been balanced in *Deinosuchus,* then we may consider the large prey species available to the huge Big Bend, Wyoming, and Montana populations. Similarly, we may ask what were the smaller (but still very big) Gulf and Atlantic crocodylians preying upon? In the case of the deinosuchid populations of Big Bend Texas and Montana, the answer seems obvious: dinosaurs of several varieties were the most likely prey. There are many dinosaur taxa in the same deposits as one finds the western *Deinosuchus* fossils (Lehman 1997), including ceratopsids (horned dinosaurs), hadrosaurs (duck-billed dinosaurs), and large and small theropods (carnivorous dinosaurs). There is, in fact, some direct evidence of crocodylian predation on dinosaurs in the Big Bend area, as will be discussed in Chapter 8. Indeed, in the sedimentary deposit containing *Deinosuchus* in the Big Bend region, there are few creatures other than dinosaurs large enough to feed the bulk of the largest *Deinosuchus*. Unlike the eastern Gulf and Atlantic Coastal Plains, the associated faunas described from the Aguja Formation in Big Bend (Rowe et al. 1992) contains relatively few mosasaurs, turtles, plesiosaurs, and other large nearshore marine prey that would seem suitable for huge-bodied carnivores. Within the limits of extrapolation from the long-dead past, the predator–prey relationship of *Deinosuchus* and dinosaurs seems firm for southwest Texas. With far less crocodylian material to evaluate, we can make the same assumption for the *D. hatcheri* occurrences in Montana. The marine beds of the Judith River Formation have been sparsely considered in the literature, but the Campanian-age Mesaverde and Bearpaw Formations in the same region contain remains of shore-dwelling dinosaurs, many taxa of fish, a very few plesiosaurs and mosasaurs, and essentially no turtles.

For the eastern deinosuchids, their generally smaller sizes leads us to suspect feeding on a more mixed bag of prey, including some smaller dinosaurs (including young theropods), many marine turtles, and probably some of the larger nearshore fish. Some of this inference comes from observations of modern alligators in habitats not too different from those *Deinosuchus* probably encountered in the southeast. *Alligator mississippiensis* is commonly observed feeding on freshwater turtles, fish, snakes, birds, beaver, and occasional deer and other larger mammals that wander into their areas. As with the Big Bend strata, direct evidence is also available to support many of the feeding assumptions for *Deinosuchus* in the east, including some impressive bite marks on several large marine turtle fossils and on at least one small dinosaur bone (Chapter 8).

But the wide diversity of prey we hypothesize for eastern *Deino-*

Figure 3.9. Deinosuchus *feeding on a giant Late Cretaceous coelacanth fish* (Megalocoealcanthus dobiei) *in southeastern United States. Drawing by Ron Hirzel.*

suchus would also make it difficult to find the evidence we might hope to use to prove our assumptions. For example, fish bones are rarely large or heavy enough to preserve bite marks. We may assume that big fish from the Campanian Age were *Deinosuchus* prey if they come from deposits with the same environmental setting in which we find *Deinosuchus* fossils, but we cannot hope to find fish bones that preserved crocodile feeding traces. As an example of a plausible but unproven prey fish species, I and others recently discovered a new genus (and single species) of very large nearshore–marine coelacanth fish, occurring in nearly all of the major localities containing *Deinosuchus rugosus* in the eastern Gulf Coastal Plain (Schwimmer et al. 1994). These coelacanth fish were heavy-bodied creatures, 3 to 4 m in length and over 250 kg in weight. They were probably slow-moving, as are the modern coelacanths, *Latimeria chalumnae* (to which the Cretaceous genus is similar—and may have been ancestral). Such fish would have been ideal prey for oceangoing, 8-m crocodylians (Fig. 3.9), and this seems like at least one likely predator–prey relationship. Unfortunately, there is no direct bone evidence to prove it.

One possible means to document *Deinosuchus* predation on fish remains is as yet untested. It would involve finding and examining coprolites (fossil feces) of the predators in hope of finding undigested remains of the prey. There is an obvious assumption that must be made in this line of research: that the coprolite and its source can be reliably connected. Fortunately, a lot of research in this subject has been pioneered with other animals, especially dinosaurs, and many of the signs used to relate animals and their coprolites are known (discussed in detail in Chapter 8). In the case of *Deinosuchus,* we would expect to find coprolites resembling larger versions of *Alligator* feces, with evidence of high concentrations of the products of bone digestion, such as calcium and phosphate salts. We might also expect to find remains of some of the more durable fish scales, such as those from the large gar *Atractosteus,* which was fairly common in contemporary deposits, and possibly the partly digested teeth of some prey land animals. I have collected many coprolites in beds rich in *Deinosuchus rugosus* fossils in Georgia, and some of these may be from the crocodylians. But so far, none of the coprolite specimens that I suspected were crocodylian in origin, and that I have sectioned and examined, has revealed identifiable bones or scales. This avenue of research remains (pun intended) a hopeful possibility.

How Could *Deinosuchus* Reach Giant Sizes?

If we accept that the giant size of many *Deinosuchus* was an adaptation for killing and feeding on large prey, we have still not sufficiently answered all the questions about their size; it is also necessary to address *how* giant size was reached. More specifically, giant animals reach their sizes by three general means: large birth sizes, rapid rates of growth, and extended periods of growth. Many large creatures experience some of all three factors, but one of these usually dominates. For example, baby elephants are born fairly large, grow rapidly during their early years, and stop their growth at maturity. This is typical of all large living mammals. In contrast, living reptiles usually hatch from relatively small eggs and grow continuously through life, but with faster rates of growth as young animals. Among larger living reptiles, the sea turtles are good examples: from eggs weighing less than a kilogram, a loggerhead or leatherback may reach 100 kg after 10 years of growth, and may ultimately weigh 300 kg at 60 years of age.

Growth in living crocodylians has been thoroughly studied in the context of zoo animals (e.g., Dowling and Brazaitis 1966) and farm-raised alligators and crocodiles nurtured for their skins and meat. It is recognized that these domestically grown crocodylians may not follow the patterns and timing of development of wild equivalents (Coulson et al. 1973), but they do allow us to make reasonable assumptions about the natural populations. In general, captive crocodylians grow rapidly for their first four to five years, continue to grow at substantial rates

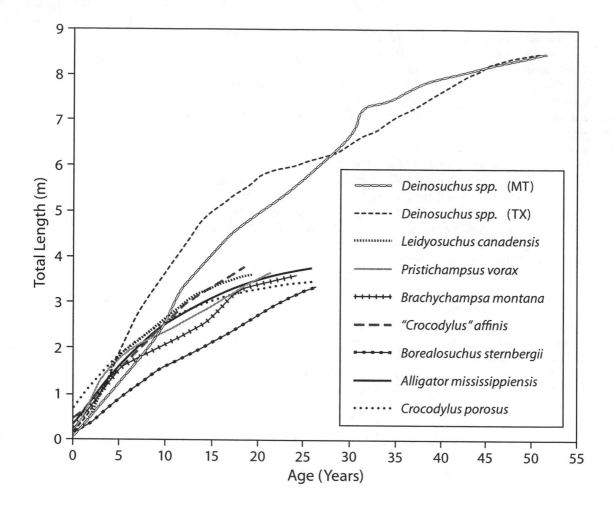

Figure 3.10. Deinosuchus *growth curve compared with other crocodylians. Reproduced and modified with permission from Erickson and Brochu (1999) (*Nature, *398: 205–206, copyright 1999, Macmillan Magazines Inc.).*

until age 10, and then experience slowing rates for the rest of their lives. For example, at age 10, a male alligator will be about 2.3 m long (Ross 1989), but at age 15, it will have only grown to a length of about 3.1 m (Dowling and Brazaitis 1966). The slight increase in length is accompanied by a substantial increase in weight, but not at the same proportional rate as the first 10 years. It is nearly impossible to measure the growth rates and longevity of individuals of wild crocodylian populations, but current ideas suggest approximately 40 years is the natural limit of longevity, and wild growth rates are comparable to those of captive crocodylians when food is abundant.

Given that we know something about the growth of modern crocodylians, what may be inferred about *Deinosuchus*? A recent report by Gregory Erickson and Christopher Brochu (1999) addressed this subject by using inference from growth rings in the dorsal osteoderms of *Deinosuchus.* Let me explain their logic: because crocodylians are cold-blooded (technically, "poikilothermic," meaning that their core body temperature fluctuates), their rate of growth and bone deposition varies with the outside temperature on a yearly cycle. Denser, thinner bone is deposited during the colder months when food is more scarce, and broader rings of less dense bone are deposited in warm seasons. Erickson and Brochu counted the rings in a diverse sample of *Deinosuchus* osteoderm specimens from Texas and Montana, as well as one that I sent to them from Georgia, and they estimated that the age of the largest animals ranged up to 51 years (Fig. 3.10), which they also assumed to represent the likely longevity of the giant crocodylians. They extrapolated from their observations on growth rings that *Deinosuchus* grew at juvenile rates (that is, relatively fast) for an extended time, up to 35 years. As a result of this prolonged growth spurt, the animals were able to reach giant size by the time the rapid growth ended (before old age) and they finally achieved truly giant sizes from this already-large midlife stage.

If this contention is correct, it would mean that the size of *Deinosuchus* was achieved differently from both giant mammals and giant dinosaurs. Dinosaurs apparently reached their size quickly, in youth, by extraordinarily fast growth (Ricqlès et al. 1998), achieving near-adult sizes by nine years of age. Mammals do approximately the same, except that they begin life as live-born neonates and need to have somewhat less relative growth achieved than did dinosaurs that hatched from eggs. Given that *Deinosuchus* was among the largest land vertebrates of all time and that its growth pattern may have been as Erickson and Brochu suggested, then we may claim that *Deinosuchus* experienced more continuous growth during its life than nearly any other known land animal. But Erickson and Brochu's results are based on the assumption that the growth rings in *Deinosuchus* represent annual seasons, as they do in modern crocodylians. If this assumption fails, then the longevity and age estimates fail too, and the growth patterns are not really known. Why should one doubt the growth-ring assumptions? One reason is that from other data, it is generally perceived that the Late Cretaceous Period, especially the Campanian Age in which *Deinosuchus* lived, was a time of relatively warm climate. Further, it is widely accepted that it was also a time of relatively less seasonality than at present: that is, there was relatively less variation from winter to summer. We may add the fact that osteoderms from specimens in southerly sites in southwest Texas and the southeastern United States seem to share the same type of growth rings as specimens living far to the north in Montana. We would expect that the southern crocodylians experienced far less seasonal change than would those in the north. One must consider the alternative explanation that the growth rings represent some other factor influencing food resources in the crocodylians, such as migrations of their prey, wet–dry seasonal climate variations, or oceanic circulation and nutrient cycles. These latter may still reflect annual cycles, but we may not be sure of this. For example, if the rings represent biennial cycles, the longevity estimate of *Deinosuchus* by Erickson and Brochu's concept would be halved, and the growth pattern and rates would come much closer to those of modern crocodylians. At present, this line of reasoning is an interesting and reasonable argument for the size of *Deinosuchus* and its growth. There are no good alternative explanations currently proposed, and the question is still wide open for study.

4. The Age of *Deinosuchus*

Dating Deposits and Fossils

In discussions so far there have been frequent references to "Late Cretaceous," "Campanian Age," and other geological time and stratigraphic terms. Many readers will understand precisely what is meant by these terms; but for those who do not, the introductory sections of this chapter include a brief explanation of the rock and time units, along with their supporting data, that are of importance in understanding the meaning of the phrase "the age of *Deinosuchus*." In the section that follows, I will detail the more precise stratigraphic and fossil associations that bear on the dating of *Deinosuchus* occurrences.

Geological dating is in many ways a curious process. It is frequently taken for granted by much of the public, which assumes that "we" know exactly how old things are. And it is about as frequently disputed and disbelieved by another, unfortunately large, part of the public, which will not accept the "Old Earth" ideas of geologists. In reality, aspects of geological dating are a highly refined science, often yielding remarkably consistent and precise dates in suitable geological situations and rock types. But in about as many situations, geological dating may be quite speculative and uncertain. Fossil deposits are commonly at the speculative end of geological dating, as will be explained below; yet it is the age of fossils that most often interests the general public when questions of ages arise. For that reason, we will examine some of the background and constraints involved in dating the sedimentary rock associations that contain fossils.

In the years between approximately 1815 and 1905, geologists actively began sorting out the patterns of sedimentary rock sequences, initially in Europe, and increasingly through the later 19th century in the United States and other diverse regions. These were still the years before techniques of absolute dating were known (that is, means to calculate actual number dates), but even earlier generations of geologists had recognized that rock strata were superimposed in time sequences, with older rock layers on the bottom and younger rock layers on top. They also realized that it took significant amounts of time for sedimentary rocks (and their fossils) to be deposited, and thus they put together the inference that stacks of rocks represented very long times of deposition. Names were given to rock units with consistent positions and characteristic rock types in the stratigraphic sequence recognized in Europe. These names often tended to reflect the nature of the rocks contained in the sequence, or alternatively, they were named for locations where their reference sections were described and measured. For example, the European Carboniferous rocks were typically full of coal, whereas the Devonian rocks were originally described in Devonshire, England, and the Jurassic rocks had their type section in the Jura Mountains. One widespread sequence of western European rocks contained great amounts of chalk, *creta* in Latin, and these became the Cretaceous rocks.

The large-scale sedimentary rock units of great thickness and areal (i.e., area coverage) extent were called systems, with the Cretaceous System as a prime example (note that systems are proper nouns, with capitalized names when they are discussed individually). Lesser units within systems were also recognized, and in descending size order they are termed series and stages (see Appendix A for a relevant list of these, with their absolute ages). It is typical for a system to have two or three series (often termed simply lower, middle, and upper), and for there to be a variable number of stages within series, usually from one to six or more. The names of stages, and some series, are typically derived from localities where the reference rock sections are best observed. This in turn creates quite a bit of complexity because separate series and stage nomenclature can (and has) been created for different continents, and even for different regions within continents. The upshot is that for the Upper Cretaceous System, we have different regional series and stages for the western United States and the European reference section. I adopt the European terminology throughout because we have to discuss time and strata in both eastern and western U.S. Upper Cretaceous rocks.

It is worth noting that the systems and their subdivisions are primarily rock units and positions within rock units, and therefore the proper terminology for the subdivisions may include the terms "upper" or "lower," but not "early" or "late." The "Upper Cretaceous," for example, must define a series, which is a rock unit, or a position in that rock unit, but it is not a date or a time unit. However, it did not require a great stretch of imagination for geologists of the later 19th century to realize that the time of deposition of the Upper Cretaceous Series must have been a significant portion of Earth time, because a great amount of sediment and many different types and levels of fossils are represented within the series. By logic, it is clear that the Upper Cretaceous Series of rocks formed during an interval we may term "Late Cretaceous" time. By these means, the original rock units were translated into time units, with a "period" representing the time of deposition of each system, an "epoch" representing the same for each series, and an "age" for each stage. In common usage, it is typical for a paleontologist to refer to the age of occurrence of a particular fossil organism in terms of ages or stages interchangeably because both units are constrained by the same types of events. That is, a rock stage usually reflects deposits forming under a regional set of conditions, bounded by changes in conditions resulting in different rock types forming. Typically, the fossils in those rocks reflect the same changing conditions, and there will usually be a new set of fossils above and below an age–stage boundary. However, rock-based units such as stages and series are not necessarily correlative across oceans because the actual rock depositional events may not be represented in all regions. Therefore, in the most precise geological communication, it is the best practice to discuss the geological ages of a particular fossil in time terms (e.g., Late Cretaceous, Campanian Age), which are global entities, free from local effects.

By use of such terms and logic, we apply so-called relative dating to a fossil or its rock setting. That is, it is implied that a given organism lived during a part of Earth history that is relatively older or younger than other times, and we may know approximately when that time occurred among the ages of Earth represented by abundantly fossiliferous rocks. For example, the Cretaceous Period is younger than the Jurassic Period, but it is older than the Tertiary Period. We know this is true because rocks and fossils of the Cretaceous Period lie above the former and below the latter wherever they are found in an undisturbed association. Where rocks with typical Cretaceous fossils are present by themselves, we may presume that the same relative positions would hold true were a fuller sequence of rocks preserved. By use of the same logic and observations, we may also delimit more precise relative associations. Thus, the Upper Cretaceous Series lies above the Lower Cretaceous and below the Lower Tertiary Series. Therefore, fossils in those rocks are bracketed between their ages. And even more precisely, within the Upper Cretaceous Series, we can recognize rocks of the Campanian Stage, which lie above those of the Santonian Stage and below those of the Maastrichtian Stage (see Appendix A for all these stages in sequence). Therefore, the Campanian Age, which happens to be the "age of *Deinosuchus,*" must be younger than the Santonian Age and older than the Maastrichtian Age. In informal discussion, scientists commonly refer to these simply as "the Campanian" and "the Maastrichtian," without specifying whether they are ages or stages, and that usage will be commonly presented here.

Absolute Dating

The spans of time represented by the various periods, series, and ages vary with the specific intervals involved because they reflect natural events rather than predetermined measures. Geological periods typically range from 50 to 100 million years in length, series range from 10 to 50 million years, and ages are usually from 1 to 10 million years in duration, all with exceptions on both ends. However, knowledge of the actual length of time of any of these time units, and their actual dates in years before the present, depends on the concept of "absolute dating."

Putting an actual numerical age on a rock or fossil is done primarily by radiometric techniques. This concept basically involves the use of naturally radioactive elements found in minerals present in the rock, with the actual dates calculated by various formulas that compare the relative amounts of radioactive parent and daughter isotopes present. The actual technique and isotopes that are used vary with the rock environment and the probable geological age involved, but they are all grounded in the assumption that decay rates of radioactive isotopes are absolutely constant over time.

However, nearly all radiometric dating is performed on igneous rocks, which ordinarily do not contain fossils—the sole exception being volcanic ash deposits, which may enclose actual fossil remains, imprints of buried bodies, or traces of past activity, such as animal tracks. Most sedimentary rock units are not directly datable by radiometric techniques for several reasons. Geologists recognize that the atomic isotopic ratios revealed by sedimentary rocks do not necessarily indicate the dates of their deposition (and therefore may not reflect the age of the fossils they contain). A sedimentary rock is very likely to show isotopic ratios that were present in the parent rock of the sediment, or it may just as well reflect several sequences of groundwater effects from events long after it was deposited. This is a common, almost universal problem in dating fossil beds.

Fortunately, there are many circumstances in which fossiliferous deposits are associated with volcanic ash beds. For example, in many sedimentary rocks in the western United States from the later Mesozoic Era (i.e., the later Jurassic and Cretaceous Periods), volcanic ash beds are interbedded with fossil-bearing sediments. This occurs because at that time, a great deal of volcanic activity accompanied early mountain-building in several parts of the Rocky Mountains, especially in Montana and Idaho. These were early episodes of the so-called Laramide Orogeny. This same process of widespread ash deposition was observed during the 1980 eruption of Mount St. Helens, when ash was spread from Washington as far eastward as Montana—and Mount St. Helens was a relatively small eruption from the perspective of geological time. The Laramide volcanoes blew ash eastward across the Great Plains, northward into Alberta and British Columbia, and southward to Kansas, causing thin ash beds to be deposited over whatever other sediment was forming in the localities immediately before specific eruptions. Because many volcanic eruptions occurred over the entire time span of the later Mesozoic, many ash beds are present. For example, in the Upper Cretaceous of Kansas, there are repeated fossiliferous chalk sequences, full of fossil fishes and a vast variety of other interesting vertebrate fossils. Interbedded with these chalks (Fig. 4.1) are a number of weathered volcanic ash beds, which have been altered to a mineral called bentonite. These bentonite ashfall layers commonly retain enough radioactive potassium to be datable by $^{40}K/^{40}Ar$ and $^{40}Ar/^{39}Ar$ decay techniques. Under such circumstances, each of the fossil beds sandwiched between the radiometrically dated volcanic layers have their

ages bracketed between the older (lower) and younger (upper) volcanics. This is one of two methods by which fossil beds may be absolutely dated, and it yields good precision when there are several volcanic ash layers to work with.

The other common method to absolutely date fossil beds applies when sedimentary rock units are cut by an igneous rock unit, called a dike. Dikes are coarse crystalline rock bodies that derive from underground magma (molten rock) that penetrated cracks in the overlying rock. They often form upright, thin, resistant rock exposures that may remain standing after the surrounding rock has eroded, and they may form a structure resembling a wall (hence the term "dike"; Fig. 4.2). Dikes tend to occur in geologically active areas where the upper crust is broken, allowing magma to intrude and cool relatively slowly; and they also frequently occur in association with volcanic areas, penetrating cracks resulting from the eruptive forces. When a fossiliferous sedimentary rock unit is penetrated by a dike, that dike may be datable by radioisotopic decay. In such cases, the age of the dike's intrusion provides a reference date that must be younger than the fossil bed that has been penetrated. If several generations of penetrating dikes are present, they may fortuitously provide multiple age boundaries. For example, if a fossil bed was deposited on one of them and then was penetrated by another, both the lower and upper boundaries of the fossil bed would be constrained. Such occurrences are known, but they are not very common.

Figure 4.1. (opposite page top) Exposure of Late Cretaceous chalk (light-colored layers) in western Kansas (Niobrara Formation), with several bentonite (altered volcanic ash, dark colored) horizons. A thick bentonite bed occurs just above the pick head. For scale, the pick is ~60 cm long.

Figure 4.2. (opposite page bottom) Igneous dike (center, vertical) cutting through volcanic ash and lava beds in Big Bend Park. Note that the dike must be younger than the center (horizontal) lava flow, through which it penetrates.

Biostratigraphy

Despite the possibilities of dating fossil beds by their associations with igneous rocks, by far most dating of fossil beds is done on a regional basis by means of associations with other fossils. This process is termed "biostratigraphy," and it might seem to be a logical circle (i.e., one fossil is dated by another, and that one by another, etc.). But this is not the case. Some fossil-bearing beds may be dated by the absolute methods described above, and these provide a large number of fixed reference points to ground the overall fossil dating in hard numbers. For example, when working with *Deinosuchus* fossils in western Georgia, we have no local or neighboring beds of same-age rock that may be dated radiometrically, but we do have a vast number of other fossil organisms in association with the crocodylian fossils. Some of these associated fossils, especially oyster shells, shark teeth, and calcareous microfossils (discussed below), are distributed widely across the Coastal Plains of the east, and also along the western side of the Interior Sea of the Late Cretaceous. I can associate the stratigraphic position of several Late Cretaceous shark species from Georgia with the same shark species in Kansas (Schwimmer et al. 1997b). And in many chalk beds in Kansas with these sharks are several interbedded volcanic ash layers (bentonites), which have been dated radiometrically; thus, the bentonites implicitly date the shark species, which in turn can constrain the age of *Deinosuchus* fossils associated with the same sharks. There are also abundant microfossils in many eastern deposits, including *Deino-*

Figure 4.3. (opposite page top) Exogyra ponderosa, a fossil oyster species that is an index fossil for the late Santonian and early Campanian, lower Blufftown Formation, west bank of the Chattahoochee River (a Deinosuchus site; see Fig. 4.7).

Figure 4.4. (opposite page bottom) Inoceramid fossils, early Campanian, Mooreville Formation, Alabama River, Jones Lock and Dam Site, central Alabama (a Deinosuchus site).

suchus beds in Georgia, for which we have elaborate global dating systems as discussed below. And further, we know from many localities in both the eastern and western United States that some fossil oysters occur in a definite vertical sequence throughout the Late Cretaceous, and we know that sequence may be related to other dating systems. By these and related assumptions, one can infer the dates of the *Deinosuchus* remains from their fossil associations, even though the actual occurrences are not directly datable.

If extinct organisms lived and died reliably everywhere at the same time, and if they were kind enough to become extinct at frequent intervals after producing similar but measurably different descendants, then biostratigraphy would be a simple and unambiguous process. Of course, this is not the case. Typically, a group of fossils (termed "index fossils") are excellent guides to a particular region's time lines (usually called biozones), but it may be difficult to correlate the index fossils from rocks of one to another region even of the same general age. For example, oysters are among the best of the larger ("macro-" or "mega-fossils") index fossils in the Upper Cretaceous eastern Coastal Plain deposits, where we also find the most abundant *Deinosuchus rugosus*. But these same oyster species (Fig. 4.3) are not found in the Western Interior in Wyoming and Montana, where specimens attributed to *Deinosuchus hatcheri* occur. Some of the same eastern oyster species are found in the Aguja Formation in Big Bend, Texas, but not in quite the same abundances and diversities as in the eastern United States. In contrast, the Western Interior Late Cretaceous deposits are full of a diverse array of ammonites (shelled cephalopod mollusks with complex sutures), as well as many types of inoceramids (large, oysterlike bivalved mollusks), where these form the basis of most western macro-fossil biostratigraphic dating systems. A few of these same ammonite and inoceramid index species are also found in both southwest Texas and the eastern United States (Fig. 4.4).

In general, for Late Cretaceous marine strata, we can do oyster biostratigraphy in the east, inoceramid and ammonite biostratigraphy in Montana, and a version of both in Big Bend and parts of the east; but can we correlate these fossil zones across the continent? Usually not, because the precise positions of biozones that are based on nearshore marine animals, such as oysters, may not be the same across great distances. For example, an animal species may appear in a given region, spread slowly to another area, and meanwhile become extinct in the first location. In such cases, the occurrence of their fossils can give asynchronous dates. Fortunately, when several fossil species co-occur in two widespread areas, they provide a much more reliable age than a single species, and we can often find multiple fossils to use in biostratigraphy. The most frequent strategy that I have used to date *Deinosuchus* remains in the eastern localities is to use biozones based on oysters and inoceramids to date sites regionally, and then to make broader correlations in marine strata by use of a variety of microfossils. As I will discuss below, many microfossil species have been correlated with absolute ages and are good age indicators themselves.

Figure 4.5. (opposite page) (top) Outcrop of the Mooreville Formation, an extensive Late Cretaceous chalk formation in western Alabama, at Harrell Station, Greene County. (bottom) Outcrop of the Late Cretaceous Niobrara Chalk Formation at Oakley, western Kansas.

But although I have sketched out how marine fossil dates can be correlated from region to region, it is often more difficult to correlate fossil deposits from marine to nonmarine settings in the same region than it is to correlate marine strata that are far apart. This nonmarine–marine problem arises especially in correlating the ages of *Deinosuchus* occurrences from the fully marine deposits of the eastern United States, with those of mixed marine and nonmarine deposits in the west (especially the Judith River Formation of Montana). In such cases, we rely especially heavily on microfossil ages. The term "microfossils" incorporates a wide range of organic remains, plant and animal, but for *Deinosuchus* work, the most important have been the plant remains commonly called "coccoliths" (short for Coccolithophoracea), also called "calcareous microfossils" or "calcareous nannofossils." These are miniscule (several microns in size) plates from many species of fossil marine chrysophyte (golden) algae, which actually comprise the substance that makes up most marine chalks. During the Jurassic and Cretaceous Periods, these tiny marine algae flourished, creating the characteristic thick chalks of those times—and coincidentally giving us powerful biostratigraphic tools.

Because of their minute size and superabundance, even a small sample of chalk will contain huge numbers of coccoliths. Further, many species of coccoliths coexisted in the plankton of the Late Cretaceous seas, raining down to the bottom sediment and accumulating throughout the epoch. A single pure Cretaceous chalk sample, or a piece of chalky shale, may contain literally dozens of different calcareous microfossil species, whose stratigraphic positions have been studied and charted globally in great detail (e.g., Sissingh 1977). In addition, these sequences of coccoliths have also been calibrated with radiometric ages (e.g., Bralower et al. 1995; Harland et al. 1990), resulting in a robust system that allows one to absolutely date a bed of sediments containing chalk. A chalk sample also usually contains a statistically huge assemblage of calcareous microfossils because they are both diverse and miniscule. For example, a single pinch of chalk, about 0.25 cm^2, from the hollow inside a tooth alveolus of an Alabama *Deinosuchus* specimen contained 10 species of calcareous microfossils. By having so many coexisting species in a single sample, these fossils mutually confirm their ages of occurrence (i.e., many species shouldn't all randomly vary in the same direction from their presumed ages of extinction). The individual coccolith species generally have age ranges averaging from about 1 to 5 million years, and assemblages of characteristic coccoliths can be delimited with average durations of 1 to 3 million years. For example, within the entire Upper Cretaceous Series, there are 15 identified sequential coccolith assemblages, and within the Campanian Stage of that series, there are six complete assemblages and part of a seventh. Therefore, on the basis of coccoliths alone—and assuming they are present in a given *Deinosuchus* sample—one may place the stratigraphic position of that fossil within a specific portion of the Campanian Stage. I have done this with at least two *Deinosuchus* remains, and the process yielded excellent precision, because of the large assemblage of coccoliths present in the specimens. And because chalk is so

valuable for geological dating, it is also fortunate that chalk makes an excellent preservational environment for vertebrate fossils. Many of the finest specimens of Late Cretaceous marine life have been collected in chalks, most famously the Niobrara Chalk sequence in western Kansas and the Upper and Lower Chalks of southwest England. But there are many other less famous Cretaceous chalk deposits with abundant vertebrate fossils, and one such deposit, the Mooreville Chalk in western Alabama (Fig. 4.5), has produced two of the most important eastern *Deinosuchus* specimens.

Microfossils other than coccoliths are frequently used to date fossil beds of the Cretaceous System, especially in sediments that are not chalky and therefore not likely to preserve calcareous microfossils. Of importance here are the extensive deposits on the Atlantic Coastal Plain, extending from New Jersey to the Carolinas. These are largely dark sands, often containing the potassium mineral glauconite, which accumulated in the bays and estuaries of the Late Cretaceous Atlantic coastline when the ocean reached farther inland than at present (as discussed in Chapter 1). Fewer carbonate-shelled microfossils are found in these deposits compared with their contemporary units on the Gulf of Mexico Coastal Plain, but there are at least some. These include species of planktonic foraminifera, which are common microorganisms closely related to amebas, but with carbonate shells (commonly called "tests"). Foraminiferan ("foram" for short) tests are important index fossils in many biostratigraphic systems, especially in the post-Jurassic, when planktonic foram species became abundant and widespread. In many cases, the forams may be correlated with coccoliths or other index fossils (e.g., Bralower et al. 1995), but during the Cretaceous, planktonic forams were not as diverse as were coccoliths. Consequently, there are far fewer assemblage zones recognized in foram biostratigraphy of the Late Cretaceous than there are coccolith assemblages. For our purposes, the Campanian, the age of *Deinosuchus,* includes either two or three recognized foram zones (depending on the author), one of which occupies most of the time span.

One additional dating technique bears mentioning: geomagnetic reversal chronology. This is a concept and technology that is currently emerging and that has proved to be of great use, especially in circumstances where other dating techniques are unusable or ambiguous. It is also a global concept and thus can help determine the big-picture date of an event. Geomagnetic reversal chronology is based on a strange phenomenon: the Earth's magnetic field reverses itself over geological time spans at quite irregular intervals. Geomagnetic reversals are a well-established fact, but the mechanism responsible for that fact is only dimly understood. Most present thinking suggests that fluid flow in the Earth's metallic core impels electrical currents that create an electromagnetic dipole, which we then perceive as the magnetic field. It seems logical that changes in these iron-based fluid currents would cause changes in the direction of the electromagnetism and hence cause magnetic reversals. It has also been alternatively suggested that core-to-lower mantle friction might cause the magnetic field, which would not significantly affect our understanding of its basic cause.

Regardless of the cause, the fact that the reversal of the Earth's magnetic field is erratic allows this effect to be used in dating. Rocks containing iron and related metals pick up traces of the magnetism that prevailed at the time they crystallize, a phenomenon termed "remnant

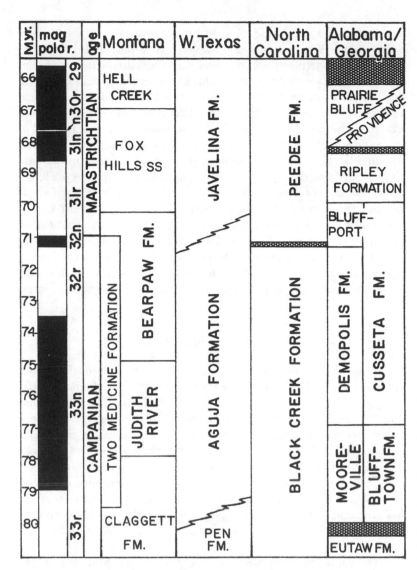

Myr.	mag polar.	age	Montana	W. Texas	North Carolina	Alabama/ Georgia
66	29	MAASTRICHTIAN	HELL CREEK	JAVELINA FM.	PEEDEE FM.	PRAIRIE BLUFF PROVIDENCE
67	30r					
68	31n n30r 31n		FOX HILLS SS			
69	31r					RIPLEY FORMATION
70						
71	32n		BEARPAW FM.			BLUFF-PORT
72	32r	CAMPANIAN	TWO MEDICINE FORMATION	AGUJA FORMATION	BLACK CREEK FORMATION	DEMOPOLIS FM. CUSSETA FM.
73						
74						
75						
76	33n		JUDITH RIVER			
77						MOORE-VILLE BLUFF-TOWN FM.
78						
79						
80	33r		CLAGGETT FM.	PEN FM.		EUTAW FM.

Figure 4.6. Portion of the Late Cretaceous paleomagnetic reversal pattern incorporating the age of Deinosuchus *(Campanian) through the end of the epoch. The stratigraphic columns are of regions with* Deinosuchus *fossils, with the dates given as the best available approximations. Data are from many sources, especially an original diagram by Dr. Tom E. Williamson, New Mexico Museum of Natural History.*

magnetism." We can easily determine the directions of remnant magnetism in a rock by use of sensitive magnetometers, and therefore we can determine if it formed during a positive or negative interval. In a vertical series of rocks either as a single stack in one locality or by correlating them, we can determine the pattern of positive and negative magnetic traces up and down the sequence. Because geologists have determined quite precisely the global pattern of magnetic reversals for the past few hundred million years, and because that pattern turns out to be erratic (i.e., with long and short positive and negative intervals interspersed), the complete effect is similar to that of a supermarket bar code (Fig. 4.6). If we can measure the geomagnetic reversal pattern in a given rock sequence, then its position within the global magnetic system is likely to be determinable.

Another important use for geomagnetic reversal is cross-checking a date that has been assumed for a given rock unit or fossil. If, for example, a particular fossil has been determined to date to a given geological age, we might be able to check magnetism in rock containing that fossil to confirm that it corresponds with the geomagnetic force field known to be prevailing at the determined age. They ought to match. In the case of *Deinosuchus* fossils, most of the time interval in which they lived corresponds with a 6-million-year geomagnetic positive interval. If we now find ambiguous crocodylian fossils in very late Cretaceous rocks that yield a negative magnetic reading, they are likely not of the age of *Deinosuchus*, and therefore probably not *Deinosuchus*. This can be of particular importance in the eastern United States for separating some postcranial remains (especially vertebrae) of the large, narrow-jawed crocodylian *Thoracosaurus* (see Chapter 7) from *Deinosuchus*, because the former persists into the Maastrichtian Age, well after *Deinosuchus* became extinct, and the Maastrichtian includes as many geomagnetic reversal intervals as it does positive intervals.

Dating *Deinosuchus*

None of the sedimentary rock beds from the exact localities containing *Deinosuchus* fossils has been dated directly by radiometric techniques—at least not so far. As explained in the preceding section, that does not preclude our having knowledge of the absolute ages of *Deinosuchus* occurrences, but it does mean that the information is indirect. Fortunately, we have several sedimentary rock units, which have been dated by radiometric means and are both geographically and stratigraphically close to *Deinosuchus* beds. In addition, we have very good biostratigraphic control on some *Deinosuchus* occurrences that are based on micro- and megafossils associated with the crocodylian bones, all of which collectively provide good age control on the age of *Deinosuchus*.

The Judith River Formation in Fergus County, Montana, which was the type locality of *Deinosuchus hatcheri* (Chapter 2), has not been dated radiometrically. But the Judith River Formation in Elk Basin, northwest Wyoming, has been dated, and in fact, it provided an important general radiometric reference point for the Late Cretaceous time scale. Hicks et al. (1995) established a $^{40}Ar/^{39}Ar$ radiometric benchmark date in Elk Basin by use of bentonites from within a thick rock section that ranges through much of the Upper Cretaceous. They dated the Judith River volcanic ash beds in Elk Basin at 79.5 ± 0.6 million years, placing this in the lower part of the Campanian Stage (see Appendix A). Further, although the Judith River Formation is nonmarine in Elk Basin, Hicks et al. were also able to correlate the argon decay dates from the Judith River rocks with the immediately underlying fossiliferous marine beds of the Claggett Shale, which has been radiometrically dated in the same area to 80.5 ± 0.15 million years (Obradovich 1993), confirming the overlying Judith River date. Because the Claggett Formation in the Elk Basin can be assigned to a specific ammonite biozone (the zone of *Baculites obtusus*), this allows the position of the Judith River Formation to be tightly correlated with respect to other marine strata in the Western Interior (including the Pierre Shale in South Dakota, which itself has been dated by argon decay; Bralower et al. 1995). The combined effect of these absolute dates and relative dates is to fix the Judith River Formation both in absolute global chronology and also in biostratigraphic position within Western Interior ammonite biozones.

In addition to providing an absolute age for a locality in the Judith River Formation, Hicks et al. (1995) also established the position of the Judith River Formation within the geomagnetic reversal system. They placed the same radiometrically dated bentonite bed in Elk Basin Wyoming at 21 m below the geomagnetic reversal boundary of C33r/C33n (translated as "chrons number 33 reverse polarity to number 33 normal polarity"). This observation provides a reference benchmark for other *Deinosuchus* beds, which have not been radiometrically dated but which may have remnant magnetism. As noted in the previous section, most *Deinosuchus* occurrences are in a positive epoch, which is C33n: the general dates attributed to this reversal interval are from ~79.3 to ~73.1 Myr (Harland et al. 1990). However, *Deinosuchus* occurrences are not precisely limited to this normal chron because they definitely range to several million years older in the eastern United States. Unlike the Judith River Formation, none of the rock units containing *Deinosuchus* fossils in the east is associated with volcanic activity or any other types of igneous rock events. They were all deposited along a tectonically passive marine shoreline, in nearshore and continental shelf settings, which were often excellent for fossil preservation but not helpful for establishing magnetic or radiometric chronology. However, some of the marine sediments in the east contain considerable amounts of glauconite (in Europe called glaucony), which is a potassium silicate mineral that crystallizes directly from marine water and is deposited extensively in marine sands. Glauconite has been used for radioisotope dating of many marine Cretaceous deposits (Harris 1982), and although the validity of the use of this nonvolcanic mineral has been questioned, it does seem to produce valid absolute ages (Craig et al. 1989). A glauconite date was derived by use of the decay of rubidium to strontium ($^{89}Rb/^{89}Sr$) from the Peedee Formation in North Carolina, yielding a date of 66.7 million years. The Peedee Formation directly overlies the Black Creek Formation, which is the source of the type specimens of *Deinosuchus rugosus*; therefore, we may at least radiometrically constrain the occurrence of the type *D. rugosus* specimens below the Peedee age.

The oldest occurrence of *Deinosuchus rugosus* documented so far is at the base of the Blufftown Formation, along the Chattahoochee River on the Georgia–Alabama border, just 15 km south of Columbus Georgia (where this is being written). A nearly complete right mandible (Fig. 4.7) was exposed on the Georgia-side bank at low river level during a particularly dry summer with the river at unusually low level. This occurrence is within a portion of the formation that has been studied extensively for other vertebrate occurrences (Schwimmer et al.

Figure 4.7. Right half Deinosuchus *mandible from the Lower Blufftown Formation, east side of the Chattahoochee River, Chattahoochee County, western Georgia (same site as Fig. 4.3). The total length of the specimen as shown is 102 cm, but there are several centimeters which were lost from the posterior end (the retroarticular process). This specimen is currently in reconstruction, and many of the gaps will be restored with fossil bone still being removed from the matrix.*

1993, 1994) and dated by the association with several index species of calcareous microfossils, shark teeth, and oysters. In precise terms, the mandible occurs in the calcareous microfossil zone of *Calculites obscurus,* at the base of the *Exogyra ponderosa* mollusk zone, and occurs above the last occurrence of the shark species *Cretoxyrhina mantelli* and with the shark species *Squalicorax kaupi.* These associations date the *Deinosuchus* jaw site slightly above the base of the Campanian Stage, which is currently dated at 83.5 million years (Obradovich 1993). I suggest an approximate age of 82 Myr for the specimen. It is important to observe that this specimen was nearly complete and obviously not "reworked"; that is, it was not deposited, eroded, and redeposited, as are many vertebrate fossils found in eastern Late Cretaceous deposits. Therefore, we may assume its occurrence provides a reliable bottom date for the genus and species.

From the stratigraphic positions of both the *Deinosuchus riograndensis* specimens in the Aguja Formation in Big Bend, Texas, and the *D. hatcheri* remains in the Judith River Formation in Montana and Wyoming, it is apparent that the genus persisted in the western United States well into the Late Campanian. The beds containing *Deinosuchus* in Big Bend have no reported radiometric dates, but they have been extensively analyzed regarding associated ammonites, oysters, and microfossils (Rowe et al. 1992). In fact, the Aguja Formation provides a useful bridge for dating Gulf Coast and Western Interior rock units because Big Bend is both geographically midway between and part of both regions. The Aguja Formation overlies the Pen Formation in Big Bend, where the Pen is a marine unit containing oysters and ammonites of early Campanian age. Above the Pen Formation are beds of the Rattlesnake Mountain Sandstone Member of the Aguja Formation, containing *Deinosuchus riograndensis.* Rowe et al. report the occurrence of three oyster species (*Crassostrea cusseta, Pycnodonte (Flemingostrea) pratti,* and *Flemingostrea subspatulata*) in the Rattlesnake Mountain beds, all of which first occur at the base of the upper Campanian in Georgia and Alabama. Putting this information together, it is apparent that in Big Bend, *Deinosuchus* does not occur as early as does *D. rugosus* in the east, appearing in the middle to upper Campanian. However, the Big Bend and eastern dates overlap considerably and do not preclude the likelihood that both regions contain the same species.

Occurrences of *Deinosuchus* in the Judith River Formation in the Western Interior are not as widespread or frequent as in the east or in

Big Bend, nor are the occurrences as well constrained by associated marine fossils. Therefore, it is not easy to trace the span of occurrence of the crocodylian through the stratigraphic sequences in Montana and Wyoming. What is apparent is that they occur in the later-early Campanian up through the middle Campanian Age. These dates also overlap with both Big Bend and eastern occurrences.

The youngest eastern occurrences of *Deinosuchus* are in the upper Black Creek Formation in North Carolina, the upper Blufftown Formation in Georgia and eastern Alabama, and the Demopolis Formation in western Alabama. These deposits date to the middle and upper Campanian, but we do not have the same dating insights available for this upper range that we infer for the lower. In part, this is logical because of the very nature of an "occurrence": if a fossil is there, it "occurs" and must have existed during a specific time. An "absence" is not as precise as an occurrence because it can represent a nondiscovery. However, assuming absences mean that the species has indeed become extinct, it appears that *Deinosuchus* became extinct near the base of the upper Campanian in the east. From my observations in Georgia and Alabama, *Deinosuchus* remains are very common until the middle Campanian, at the boundary between two formations (the Blufftown and Cusseta Formations), and then they are not found above that horizon. The last reported occurrences of *Deinosuchus* in Mississippi are in the Demopolis Formation (Manning and Dockery 1992), at a time horizon I believe is equivalent to the age of the Blufftown–Cusseta Formation boundary where it occurs in Georgia. This time would be close to 77 million years ago. Because the Mississippi occurrence is the westernmost of the "eastern" sites for *Deinosuchus* (i.e., on the eastern side of the Western Interior Seaway), and because the Georgia occurrence is the easternmost site in the Gulf of Mexico, this suggests that *Deinosuchus* across the southeastern United States became extinct at 77 million years ago, which is slightly earlier than the last *Deinosuchus* occurrences in the Western Interior and Texas.

Before leaving this subject, I should comment on occasional reports of *Deinosuchus* fossils in rocks younger than the Campanian, especially on the Atlantic Coastal Plain. I have even observed (especially in Internet discussions) that *Deinosuchus* is sometimes listed as one of the large animals that allegedly survived the mass extinction of the Cretaceous–Tertiary (K-T) boundary. Although I have not seen published comments or formal reports on this subject, I have been told by amateur and some professional sources that late-occurring *Deinosuchus* remains are especially well known in the uppermost Late Cretaceous (Maastrichtian) formations in New Jersey and Delaware. Most of these verbal reports are based on discoveries of very large vertebrae of the advanced crocodylian type termed "procoelous," which is characteristic of *Deinosuchus* and all eusuchian crocodylians (see Chapter 7). In addition, large crocodylian teeth are also reported to occur in the same strata (although none reportedly shows the rounded morphology of the rear teeth of *Deinosuchus,* which is noteworthy). I deduce that all of these reports are incorrectly associating very large crocodylians with *Deinosuchus* and that the most likely source of these Maastrichtian

Age fossils in the North Atlantic region is *Thoracosaurus neocesariensis*. This species was a widely distributed, long-snouted crocodylian that seems to have ranged from the Campanian Age well up into the Early Tertiary Period. It may have been a basal ancestor of the "false gavials" in the modern genus *Tomistoma* (see Chapter 7). I was especially impressed to discover the size this common species could reach when I observed a skull from the Maastrichtian of New Jersey, in preparation by Dr. Barbara Grandstaff at the New Jersey State Museum in Trenton. The skull was at least a meter long and fairly massive, suggesting that the crocodylian was approximately 7.0 to 8.0 m long and may have weighed 2.0 tons. The vertebrae of such an animal would indeed resemble those of *Deinosuchus* in size and shape, and the front teeth would be generally similar on cursory inspection (but the posterior teeth would be slender and pointed). I conclude that this species is the Maastrichtian and era-boundary-crossing big crocodylian, rather than *Deinosuchus rugosus*.

5. *Deinosuchus* Localities and Their Ancient Environments

Atlantic Coastal Plain

Throughout the text to this point, we have collectively examined the *Deinosuchus* occurrences spread across the United States, from New Jersey to Montana. In Chapter 6, I will analyze the evidence bearing on the question of whether these widespread locations all harbored the same species of *Deinosuchus*. But first, in this chapter, I will examine in detail the information available about the ancient environments represented by the various *Deinosuchus* fossil localities. As will be discussed, fossil deposits with *Deinosuchus* were formed in different conditions and settings across the various regions and sites, and the fact of this habitat diversity suggests that the crocodylians had a wide range of tolerances. A complete list of known *Deinosuchus* localities and their geological data is presented in Appendix C for ready reference. The following discussion will group the fossil localities by regions and other categories.

To begin in the eastern United States, we may divide *Deinosuchus* localities into the Atlantic Coastal Plain and eastern Gulf Coastal Plain regions (Fig. 1.3). These large areas form naturally distinct subdivisions when considering fossil deposits of the Late Cretaceous because there are no intervening fossiliferous sites of that age from southern South Carolina through central Georgia. Therefore, the deposits from New Jersey south to northern South Carolina constitute the Atlantic region, and deposits from western Georgia to Mississippi are the eastern Gulf region. In general, the Atlantic deposits are more uniform than are those in the eastern Gulf, and we shall examine them first.

The noteworthy *Deinosuchus* beds on the Atlantic Coastal Plain are spread widely apart. In New Jersey, there are several sites in the Marshalltown Formation of Monmouth County that have produced good *Deinosuchus* material, and the most intensively studied and described is at Ellisdale (Gallagher 1993), which we will focus on here. In North Carolina, there is Phoebus Landing and neighboring sites in the Black Creek Formation of Bladen, Sampson and Duplin Counties, which have been discussed in Chapter 2 because these collectively form the type locality of *Deinosuchus rugosus*. These two general localities are not at all the only sources of *Deinosuchus* fossils along the Atlantic Coastal Plain, but they seem to contain especially large populations, and there probably were good reasons why this is so. One significant factor that bears on the amount of *Deinosuchus* fossil known from these regions and specific sites is the amount of study they have received. These two localities have been extensively studied by generations of paleontologists, whereas much of the remaining Atlantic Late Cretaceous deposits of Campanian Age (the "age of *Deinosuchus*") have received much more scanty professional attention. Fossil collectors along many spots of the Atlantic Coastal Plain outcrop find *Deinosuchus* teeth and osteoderms at many localities but do not formally describe them. For example, *Deinosuchus* remains are fairly common in the Chesapeake and Delaware Canal exposures in Delaware, in the Marshalltown Formation. This deposit has been extensively collected for its fossil fishes and marine reptiles (e.g., Lauginiger 1984, 1988), but the crocodylians (including *Deinosuchus and Thoracosaurus*) have only been mentioned in passing as occurrences. I have also seen extensive amateur collections from two sites in the Black Creek Formation in South Carolina (Burches Ferry in Florence County and Kingstree in Williamsburg County), which contain *Deinosuchus* teeth and other bones. But these have never been described other than anecdotally (as I am doing here; see also Weishampel and Young 1996 regarding the dinosaurs from these sites). We may assume that *Deinosuchus* was a common presence across the entire Atlantic coast during the Campanian Age of the Late Cretaceous. But the two localities to be highlighted—Ellisdale and Phoebus Landing—seem to have extraordinary concentrations of the beasts, and there is a definable reason for these concentrations, as we shall see.

The Ellisdale Locality has been intensely studied during the past decade by a group of paleontologists working out of the New Jersey State Museum (Gallagher 1993; Parris et al. 1987). Like most Atlantic Coastal Plain Cretaceous fossil localities, Ellisdale is not physically impressive to the observer. In the east, fossil beds tend to be grassed over and are frequently located in forest, swamp, or stream beds—after all, modern environments have no necessary relationships with the conditions prevailing during the Late Cretaceous, and a former seacoast (or desert) of the Mesozoic may be today's forest (or shopping mall!). At Ellisdale, the bed producing the vertebrate fossils is a thin layer of dark sand and flat pebbles close to the level of a small stream; and if one looks closely through the gloomy overhang of trees, one sees a lot of isolated, black-stained bones among the sediment. This bed is found in a wooded area, accessible along the stream, and in proportion to the very small size of the exposure, it has been visited by many vertebrate paleontologists and amateur collectors. The origin and nature of the deposit is revealed by considering the type of sediment it contains, the mix of fossils within, and the nature of the fossil preservation.

The dark, silty sands of the Marshalltown Formation at Ellisdale contain a lot of glauconite and some mud, and the overall sedimentary deposit lacks level bedding surfaces. Collectively, these are clear indications that the formation accumulated as a nearshore marine deposit, close to the mouth of one or more large rivers. This same type of sedimentary deposit may be found in modern bays and lagoons along the Atlantic coast, as well as many other temperate-climate coastlines. To a paleontologist, the mix of fossils contained in a sedimentary deposit such as the Marshalltown is just as revealing of its conditions during formation as is the sediment. Another good source of information is the physical condition of the fossils found at Ellisdale (and other major eastern *Deinosuchus* sites); by condition, I refer to both their state of wear and disarticulation, and of mineralization. Among the fossils at Ellisdale, we find a striking mixture of marine and nonmarine species, both invertebrate and vertebrate. On a statistical basis, the bias is strongly toward marine species, but the nonmarine portion is too large to be ignored. For example, at Ellisdale, there are remains of many types of turtles, including families that are generally nonmarine in modern times (e.g., the soft-shelled trionychids) and those that are unequivocally marine (e.g., the cheloniid and protostegid sea turtles, and the pelomedusid side-necked turtles). There are also many dinosaur fossils, a few lizard remains, and even some rare mammal teeth, along with a considerable amount of fossil wood and amber. All of these components point toward a large nonmarine input to the deposit.

However, the invertebrates in the fossil assemblage, especially the clams and snails, are exclusively marine species or those with wide salinity tolerances (such as oysters). Also, notwithstanding the dinosaurs and such, the majority of vertebrate fossils come from marine animals, including many types of sharks and rays, bony fishes, mosasaurs (very large marine lizards), and a single species of plesiosaur. There is also a long list of crocodylians present at Ellisdale (and other New Jersey Marshalltown sites) in addition to *Deinosuchus*. Because our purpose here is evaluating the paleoenvironment of *Deinosuchus*, we obviously can't use its presence at Ellisdale as an indicator of habitat; but we have other species present with better-known preferences. Among the likely nonmarine genera at Ellisdale are several smaller alligatoroids (see Chapter 7), including *Brachychampsa* and *Allognathosuchus*. The first has been found in obviously nonmarine sites in the Western Interior, whereas the habitat of *Allognathosuchus* is not as clearly demarcated. The other crocodylians, including *Leidyosuchus* and *Thoracosaurus*, may have favored a wide variety of habitats and need further evaluation, as we are doing with *Deinosuchus*.

To pinpoint and interpret the paleoenvironment represented by Ellisdale and similar sites, we must apply the principle of Uniformitarianism ("the present is the key to the past") discussed in every historical geology textbook. This means finding a modern analog for a past condition, and then correlating the one to the other. For the set of fossils and sediments at Ellisdale and other eastern *Deinosuchus* sites to be discussed, we find our modern analog in larger Atlantic tidal estuaries. These are nearshore areas at the mouths of coastal rivers, where seawater and river water interchange with the tides. Gallagher (1993) discussed this idea in some detail when considering the paleoenvironment at Ellisdale, and I presented a similar interpretation for the paleoenvironment of another major *Deinosuchus* site, at Hannahatchee Creek in Georgia (Schwimmer 1997b). Indeed, the three most important *Deinosuchus* sites in the eastern United States (Ellisdale, Phoebus Landing, and Hannahatchee Creek) share in common so many paleoenvironmental characteristics that they almost certainly represent similar origins. And where they differ in specific details, such as the presence or absence of a particular species, there is usually an analogous species that appears to fill the same ecological niche in the fossil assemblage.

More evidence that Ellisdale (as well as Phoebus Landing and Hannahatchee Creek) represents a fossil estuary, besides the nature of the sediment and the fossil mix, is the abundance of clam borings in wood, shells, and sometimes bones. In these fossil deposits, it is typical for larger mollusk shells and fossil wood (which is usually altered to lignite or subbituminous coal) to be riddled with holes (Fig. 5.1). The larger holes, typically from 5 to 12 mm in diameter, were bored by members of two groups of bivalves, the Pholadids (which include the "shipworm," *Teredo*), and *Lithophagia* (whose name means "rock-eater"). In modern seas, these same types of clams are observed to exist in estuarine and brackish waters, and their borings in the Cretaceous fossils strongly imply that the bored shells and wood lingered in the same types of settings. (I use the term "lingered" here because the depositional processes evident in these tidal estuaries also may include transportation of materials away from their original sites of deposition.)

We also find a variety of other types of holes bored in shells and bones at many Late Cretaceous localities, which may be attributed to additional types of organisms. Among these other borings, I attribute small holes on shells and bones (typically 2–4 mm diameter) to clionid sponges, especially where those holes are very abundant and blind-ended (i.e., they do not pierce the shell). In some sites, nearly all of the mollusk shells have been bored by clionids, often on both exterior and interior surfaces; this indicates that the mollusk was dead before the boring was completed, and it implies that the shell was rolled around on the sea bottom (and likely transported). Similar-sized holes on clam-shells, which are almost always singular on a particular shell, may be attributed to drilling gastropods such as moon snails. Fairly recently (Kase et al. 1998), it has been argued that sets of larger holes (typically >1.0 cm) in Late Cretaceous ammonites may be attributed to limpets, which are another type of snail. However, this contention is a matter of dispute by paleontologists who accept the traditional explanation that these holes represent predatory bite marks by mosasaurs (G. Westermann and C. Tsujita, personal communication). Resolution of this dispute is an important consideration for our purposes because we must be aware of the alternative explanation of invertebrate borings when eval-

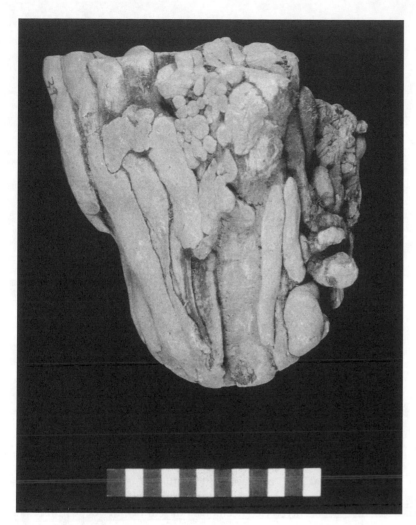

*Figure 5.1. Clam-bored (*Teredo sp.*) fossil wood, which has been preserved as subbituminous coal. From the Blufftown Formation, early Campanian, Hannahatchee Creek, Stewart County, Georgia. Scale in centimeters. Photograph by Tracy Hall.*

uating alleged *Deinosuchus* feeding traces in bones (Chapter 8). However, despite the variety of borings that may occur in Late Cretaceous fossils, the clam borings are distinctive and easily distinguished from the other types discussed here, which are found only in fully marine habitats. Therefore, the presence of clam borings at Ellisdale and other sites points to their occurrence in brackish waters, which confirms the estuarine model.

The North Carolina localities centered on Phoebus Landing along the Cape Fear River (Fig. 1.4) have many geological characteristics similar to those at Ellisdale, suggesting that these too represent one or more Atlantic coastal estuaries during the Campanian Age. The Phoebus Landing site has been extensively studied for many years (Baird and Horner 1979; Emmons 1858; Miller 1967, 1968; Weishampel and Young 1996), even though it lies along a section of the river with very difficult access. In fact, the fossil beds are often underwater and can then only be reached by boat. The fossil assemblage there, like Ellisdale, is a scrap fauna of isolated bones and teeth, with nearly the same mix of fossils, including several types of dinosaurs, freshwater and marine fishes and turtles, mosasaurs, and crocodylians, including abundant *Deinosuchus*.

It is worth considering the taphonomic (i.e., the fossil preservational) histories of these Cretaceous scrap faunas found at sites such as Phoebus Landing, Ellisdale, and Hannahatchee Creek in Georgia (discussion of which will follow). As I have explained, the mix of fossil species and observations such as clam-bored wood and bones tells us the living material accumulated in an estuary. But details about the fossils and their accumulations also tells us something else: that these deposits have been reworked in the ocean and concentrated, very likely offshore from where they originally collected. The clues to this information are in the high percentages of vertebrate materials present and in the details of individual bone and tooth preservations. If one strolls along a modern beach or strandline, looking for living shark teeth, it would be good luck to find just one tooth in a whole day's search; that's the normal concentration of teeth on a typical shoreface. (However, fossil shark teeth are frequently eroded up from sediment under modern shores, so this experiment must be done carefully.) In contrast, many Cretaceous scrap faunas will produce dozens of fossil shark teeth on a surface of 1 m^2, and some extraordinary localities are composed of nearly pure vertebrate teeth and bone (Fig. 5.2). Clearly, these types of deposit do not represent normal accumulations. Also, in most high-concentration coastal deposits, close examination reveals that the teeth of reptiles, bony fish, and sharks are mostly blunt, rounded, and broken, as are virtually all bones and most other vertebrate fossils (but often not mollusk shells—another clue). Then too, in these same deposits, bones are rarely, if ever, found in mutual association; that is, skeletons are disassembled, and bones are nearly always isolated.

What has occurred is a mechanism of sedimentary deposition that has become elucidated in the past two decades as part of a larger model of "sequence stratigraphy" (Haq et al. 1987). Sedimentary geologists now realize that fluctuating, cyclical sea levels have dominated the history of nearshore areas of continents for hundreds of millions of years, possibly much longer, and that rising and falling sea levels cause distinct, recognizable depositional patterns. One of the common cyclical processes involves the effects of marine transgressions (i.e., seawater encroachments) onto the continental shelf and shores. During part of these transgressions, vertebrate bones and teeth, which accumulate on the deeper subtidal continental shelf, are transported upward into the shallower marine zones from deeper waters. As they are transported and redeposited, they become concentrated as a result of several processes. The first cause results from vertebrate material being relatively dense compared with the carbonate shells and silicate sands of typical marine deposits. Being denser, it tends to drop out and accumulate in concentrations within every depression and surface irregularity on the sea bottom. These deposits, called "lag concentrations," form in many of the eastern coastal vertebrate bone beds. This same lag concentration effect is frequently used by fossil collectors to find phosphatic fossils in

modern streams; these collectors search natural crevices as catch areas for fossils washing in from the banks and upstream sites; it is generally similar to the strategy behind panning for gold.

Another mechanism of concentration associated with marine transgressions relates to "sediment starvation." As part of the sequence stratigraphic model, it is predicted that sea level cycles will include episodes when there is a limited amount of material available to deposit on the newly transgressed surfaces. These sediment starvation events occur when sources of sand and silt from rivers are cut off from the shorelines. Under such conditions, the vertebrate material washed in from offshore by the transgressing ocean waters becomes a major sedimentary component and is thus highly concentrated. Such marine deposits are termed "condensed sections" in the jargon of sequence stratigraphy, and they form another major type of eastern Cretaceous bone bed. It is likely that the bone-rich deposits at Ellisdale and Phoebus Landing are condensed sections with complex histories. I believe the bones and teeth originally accumulated in estuaries where the animals lived and died. They were later transported into deeper waters of the ocean during a regressive phase of a sea level cycle, then were subsequently retransported back high up onto the marine shelf during a transgressive phase. Each sequence of transportation caused further rounding and abrasion of the vertebrate material, with breakage of bones, and increasing concentration of the vertebrate portion of the sediment. Such a complicated explanation might seem improbable, ex-

Figure 5.2. Fossil tooth concentration in a lag deposit, upper Blufftown Formation, Barbour County, Alabama. All the darker material is composed of water-worn shark teeth, mostly Scapanorhynchus texanus. *Photograph by Tracy Hall.*

cept that the concept of cyclical sea levels fits well into the back-and-forth transportation idea.

It is remarkable to realize that the most highly concentrated deposits are likely to have undergone more than one cycle of transportation: that is, they were concentrated and deposited as described, then later eroded, retransported, and redeposited even later in time. By this reasoning, the final time of deposition of such beds may have been much later in geological time than the age in which the actual animals lived. This redeposition and transportation concept also means that the localities of final deposition may not be where the organisms originally lived; that is, the *Deinosuchus* fossils at Phoebus Landing and other sites may come from crocodylians that actually lived many kilometers up- or downshore from the site (but almost certainly they traveled by upstream ocean currents, from whichever way that was at the time). As a final note of caution, we must also realize that postmortem transportation also allowed for mixtures of fossils from different places, and it is very possible that the fossils deposited at one of these *Deinosuchus* beds may come from more than one original site.

Nevertheless, despite evidence that the *Deinosuchus* sites discussed here were obviously concentrated and reworked, my impression in the field is that however much in-and-out marine shelf and shore transportation may have been involved, there probably has not been an appreciable amount of lateral transportation in these localities. I base this assumption on my observation that the same general mixture of fossils persisted in the same region through the entire early Campanian. It seems improbable that all of these same species would be repeatedly transported together from the same distant sources. In other words, they may have been washed down and up the continental shelf, but not necessarily along the shoreline. Therefore, I believe that the mix of fossils at Phoebus Landing and related sites probably lived in close proximity to each other and to the final depositional locations.

Eastern Gulf of Mexico Coastal Plain

Turning the eastern continental corner, so to speak, around southern South Carolina, Florida, and most of Georgia (i.e., those areas where there are either no exposed Cretaceous strata, or none that would likely preserve vertebrate fossils), we make the transition from the Atlantic to the eastern Gulf Coastal Plain. During the Late Cretaceous, this continental corner was largely under seawater; there are Late Cretaceous marine deposits buried underground, but they are only detected in boreholes drilled for petroleum exploration. The easternmost *Deinosuchus* beds, which are technically classified as being in the Gulf Coastal Plain, are found in western Georgia, nearly astride the border with Alabama, in the valley of the Chattahoochee River. It is in these beds that I began collecting and studying *Deinosuchus* fossils, and I believe they still contain the highest concentration of the giant crocodylian's bones and teeth of any known locality. Many sites in the Chattahoochee River Valley have yielded *Deinosuchus* remains, but the

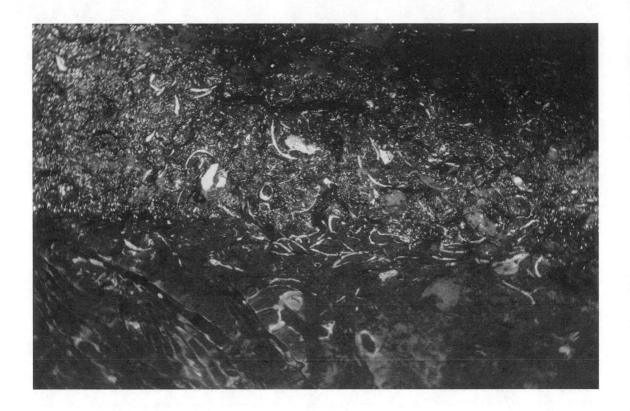

foremost collections come from the banks of Hannahatchee Creek, Stewart County, Georgia.

Along Hannahatchee Creek are several kilometers of bank out-crops that expose the upper Blufftown and lower Cusseta Formations, and especially the contact zone between the formations. This thin zone of formation contact, less than a meter thick, contains a concentrated mass of shell and bone scraps (Fig. 5.3), many of them clam-bored (as at Ellisdale), indicating that it accumulated in brackish water and was redeposited during a marine transgression. I have studied, dated, and worked with this very rich fossil concentration horizon for many years, and the general nature of the site has been described by me in two semitechnical articles (Schwimmer 1986, 1997b) and with several col-leagues in technical articles about the dinosaur fauna (Schwimmer et al. 1993) and the fishes (Case and Schwimmer 1988).

The overall aspect of the sedimentation and fossil assemblage in the upper Blufftown Formation at Hannahatchee Creek most closely resemble conditions in the Black Creek Formation at Phoebus Landing. The noteworthy differences observed at Hannahatchee Creek seem to be that there are many more fish remains present, both in absolute numbers and diversity, many more marine turtle individuals present (but of only a few species), relatively fewer dinosaurs, and extraordi-narily abundant *Deinosuchus* teeth. In fact, this locality so far has the highest concentration of *Deinosuchus* fossils known in any fossil site.

Figure 5.3. Fossil shell and bone concentration zone caused by marine transgression, in the upper Blufftown Formation, middle Campanian age, Hannahatchee Creek, Stewart County, Georgia. The white material is bivalve shells; the dark, irregular band of material near top center is coalified wood.

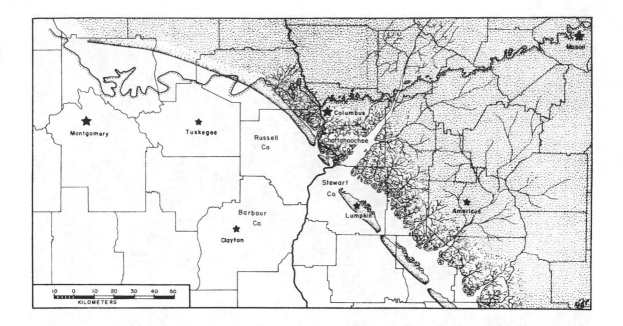

Figure 5.4. Reconstruction of the paleoenvironment of the southern coast of western Georgia and eastern Alabama, during the middle Campanian. The large estuary in the center of the figure is the location of the Hannahatchee Creek site. Other notable Deinosuchus sites in the figure are in eastern Russell County, Alabama, in the center of the figure. Drawing by W. J. Frazier.

There is also plenty of *Teredo*-bored fossil wood present, as well as freshwater turtle and nonmarine crocodylian fossils (mostly *Leidyosuchus*) in the assemblage, to show that Hannahatchee Creek locality was almost certainly located within a tidal estuary during the Campanian Age (Fig. 5.4). However, the subtle differences from the Phoebus Landing fauna suggest that the western Georgia site was slightly more influenced by marine conditions than was the North Carolina site, and thus perhaps it formed closer to the open ocean. For example, it contains many large ammonites (shelled marine cephalopods), whose shells float after death and tend to be found more commonly in settings open to the sea than in those with restricted access. But like the other eastern sites discussed so far, at Hannahatchee Creek, *Deinosuchus* bones are always isolated specimens (i.e., they are never found articulated and rarely intact).

One of the most significant discoveries at the Georgia site is abundant evidence that *Deinosuchus rugosus* fed frequently on marine turtles (discussed in Chapter 8). This is based on various types of evidence, but especially on the discovery of several turtle fossils with large, blunt bite marks that neatly fit posterior *Deinosuchus* teeth. One factor present at Hannahatchee Creek that may have influenced our discovery of crocodylian feeding traces is the high quality of bone preservation at the site. Even though the fossil bones are isolated and water-worn, they are hard, uncrushed, and well preserved by black, phosphatic mineralization. This high-quality surface preservation allowed even subtle feeding traces to be preserved, including scrapes left by small sharks scavenging several types of large animals, including turtles and dinosaurs (Schwimmer et al. 1997b). The crocodylian feeding signs are large and evident, and the quality of their preservation allows us to confirm that they were caused by crushing forces rather than by boring activities (see Chapter 8).

Continuing the survey of *Deinosuchus* sites westward across the Gulf Coast, we come to several that have yielded only one important *Deinosuchus* individual at each site, and none that routinely includes many *Deinosuchus* fossils. It is noteworthy that the specimens from the central areas of the Gulf Coastal Plain (which I define here as central Alabama to Mississippi) tend to be sets of associated remains or partial skulls and jaws, rather than the isolated bones we find farther east. In other words, westward into central Alabama, yet still within the eastern Gulf Coast, we find far fewer individual *Deinosuchus* fossils, but those we do find include much more complete specimens than we have noted farther to the east. There are two possible explanations for this effect, both based on the fact that the Campanian-age deposits toward the central Gulf coastal area formed on the open marine continental shelf, rather than in nearshore environments, as on the Atlantic coast and easternmost Gulf coast.

Deeper continental shelf deposits of thc Late Cretaceous are typically composed of chalk, often quite pure but sometimes mixed with a great deal of clay, or with volcanic ash in the west. The deep-water environments that received mostly chalk sediment were subject to much less disturbance than were nearshore deposits because they formed below wave base levels and were rarely reached by storm surges. Thus, vertebrate carcasses deposited in chalk sediments had much higher probabilities of lying undisturbed long enough to be buried, with many bones preserved in articulation. Chalk sediment also preserves bone well, with the only problem being the tendency for chalk fossils to be flattened when the chalk dewatered and compressed as it turned to rock. As an added bonus, chalk is itself composed largely of coccolith fossils (see previous discussion on dating) and usually allows very precise age assignments for the fossils that it encloses. In fact, we might expect better preservation of many large vertebrate fossils in Late Cretaceous chalks than we actually do observe, except that there is also evidence that many carcasses were scavenged on the sea bottom by several types of sharks (Schwimmer 1997b; Schwimmer et al. 1997b). This shark scavenging, especially by species of the serrate-toothed genus *Squalicorax*, apparently caused a lot of bone scattering. (It is easy to recognize when *Squalicorax* was the scavenger because it was the only Late Cretaceous shark genus with serrate teeth, which left clear signs in scavenged bones.)

The best explanation for the relatively rare occurrence of *Deinosuchus* in the deeper marine deposits of the central Gulf coast may simply be that the animal did not frequent deep waters. The few specimens that have been found may have been transported to these environments after the individuals died. We do know that occasional dinosaur carcasses are found in marine chalks and other marine deposits (e.g., Langston 1960; Schwimmer 1997b; Schwimmer et al. 1993), and they certainly weren't derived from animals swimming in the deep ocean. Therefore, it is clearly established that nonoceangoing animals can be transported and preserved in deep marine environments; however, this

Figure 5.5. Chalk surface in the Mooreville Formation, Jones Dam site, Lowndes County, Alabama. The fossil in the center of the frame is an inoceramid (cf. Fig. 4.4).

begs the question of whether or not *Deinosuchus rugosus* was truly an oceangoing crocodylian, like the modern saltwater *Crocodylus porosus*. In Chapter 8, we will consider evidence that *Deinosuchus* fed on marine turtles, with one well-bitten marine turtle specimen, found in western Alabama chalk, adding some weight to the idea that *Deinosuchus* did venture into deep water. But overall, we do not see nearly as much *Deinosuchus* sign in deposits from offshore settings as we do from the nearshore. Because the more westerly Gulf coastal exposures of the Late Cretaceous formed in relatively deeper ocean settings, we are fortunate to have at least several *Deinosuchus* sites filling in a big geographic gap between eastern Alabama and west Texas.

A single significant *Deinosuchus* discovery from central-western Alabama comes from the Mooreville Chalk in western Lowndes County. Lowndes County is located on the approximate eastern edge of the western third of the state, and the site is on the north bank of the Alabama River near an Army Corps of Engineer's dam (Fig. 5.5). The specimen probably consisted of the front half of a skull and half a complete lower jaw (the right ramus). The "probably" in the previous sentence reflects the interesting history of this *Deinosuchus* discovery, in which I had a belated part.

The specimen was literally stumbled upon by a boatload of fisherman who landed on the riverbank, apparently to urinate collectively, when they saw the fossil exposed on the bank slope. It must have been quite a strange sight to the fishermen: a meter-long tan- and brown-stained slab with huge teeth standing above the gray chalk sloping down to the water. They believed it was the skull a giant fish (I was told it was thought to be a gar), and they attempted recovery of the specimen as best as they could. Unfortunately, at some point they also apparently proceeded to remove the huge premaxillary teeth by breaking up the anterior bones. They may have done the same to the lower jaw (I was told about the damage to the premaxilla, and I deduced the lower jaw damage). From what may have been a weathered but nearly intact field specimen, the fossil was converted into literally hundreds of pieces of fossil bone and teeth, plus about 10 kg of fragments and chalk dust. The large premaxillary teeth disappeared, and I learned indirectly that they crumbled apart several months after the specimen was found, probably because of the destructive effect of untreated "pyrite disease." Pyrite disease is due to the spontaneous growth of iron sulfide crystals (the mineral pyrite), which occurs in wet sedimentary rock and fossils that are buried in conditions that are rich in both sulfur and iron. When pyritized fossils are dried and then reexposed to atmospheric humidity, the pyrite crystals degrade and release sulfuric acids that eat away at fossil bone and teeth. Pyrite disease can totally destroy an unprotected specimen and is a serious problem in museum collections, especially with specimens from the eastern United States because of the humidity. It can be controlled by isolating the specimen from humidity by use of various plastic resin coatings, and it may be stopped by proprietary chemical fumigants.

Unfortunately, the Alabama fishermen were unaware of this problem, and without benefit of more than a year's pyrite disease control, and after complex transactions, the *Deinosuchus* material arrived at my lab on a slab of plywood. I identified it instantly by the teeth and bone texture and spent parts of the next three years curating and reconstructing the specimen (Fig. 5.6). Even though it suffered greatly in its exposure and handling, the restored specimen is still one of the most informative eastern *Deinosuchus* fossils available for study. It also provides the front half of the computer-spliced reconstruction presented in Figures 1.7 and 1.8.

Another very important specimen came from the marine chalks of far western Alabama, collected near West Greene, Greene County. The specimen contains the most complete postcranial skeleton of a single *Deinosuchus* known in the eastern United States (and possibly from anywhere), including 29 vertebrae from most body regions, most of the hips, a femur, several hand and foot bones (metapodials), and, notably, approximately 50 osteoderms. It completely lacks any part of the skull or mandible, but the osteoderms are undoubtedly those of *Deino-suchus,* and the identification is firm. Nevertheless, the specimen is very small for the genus, with the femur and ilia of the hip corresponding to a large modern alligator in general size (~ 4.0 m). It is evidently a young individual, judging by the size and slenderness of bones; and according to the growth curves of Erickson and Brochu (1999), a *Deinosuchus* of this length was approximately 12 to 15 years old at death.

Figure 5.6. Ventral view of the anterior two thirds of the Deinosuchus skull from the Mooreville Formation, Jones Dams Site, Lowndes County Alabama. Reconstruction of the specimen is still in progress because of the extreme damage, especially to the crucial region at the junction of the nasals, maxillae, and premaxillae.

The history of this specimen is tortuous. According to Thurmond and Jones (1981), it was originally collected during the 1950s by a troop of Boy Scouts and deposited in a private house until one of those two authors (unspecified) recovered the specimen for study. (The rock formation of origin was not recorded, but it appeared to me typical of the pale chalk of the early Campanian Mooreville Formation.) The specimen was cataloged in the Alabama State Museum collections and was later examined by several workers (including Dr. Wann Langston

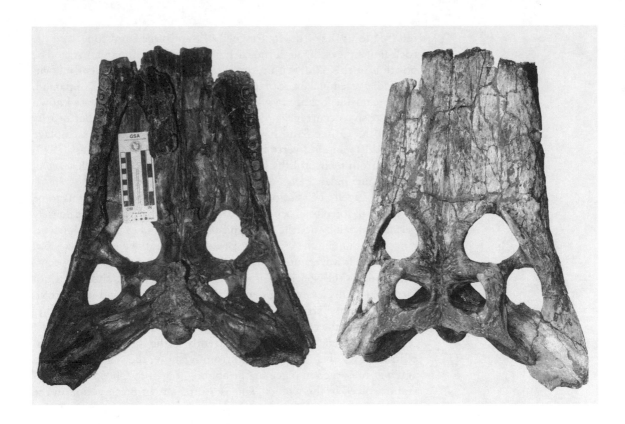

Figure 5.7. (left) Ventral and (right) dorsal views of the posterior Deinosuchus skull unit from the Coffee Sand Formation, Lee County, Mississippi. Ventral view photograph by D.R.S.; dorsal view photograph courtesy of Wann Langston Jr.

Jr. and Dr. Samuel W. Shannon of the Geological Survey of Alabama), and the identification was made and confirmed. At the time I briefly examined the specimen, it was under study by James P. Lamb, a doctoral candidate at North Carolina State University in Raleigh. To date, no formal description of this specimen has been made, but my cursory examination showed that the body was nearly identical to a modern alligator of comparable size. Indeed, the ilium (which I was particularly interested in because I had a comparable-sized *Deinosuchus* ilium from Georgia in my laboratory) was identical to that of *Alligator*. The presence of this headless, juvenile *Deinosuchus* in open-sea chalk deposits is suggestive that this was a true marine occurrence of the crocodylians. But as I have explained, even dinosaurs have been found in essentially the same deposits (Langston 1960), and we cannot make too many conclusions about this single occurrence in western Alabama chalk.

The westernmost occurrence of *Deinosuchus* on the eastern side of the Mississippi Embayment of the Gulf Coastal Plain (see Fig. 1.3) consists of a single well-preserved posterior half skull (Fig. 5.7), a fragment of mandible, and several isolated teeth and osteoderms from northeastern Mississippi. It is, in fact, the posterior skull that is the best-preserved specimen known for the genus and served as the rear portion of the reconstruction in Figures 1.7 and 1.8. The specimen comes from the Coffee Sand Formation, along Tulip Creek, in the vicinity of Tupelo, Lee County, Mississippi. This is a rock unit often referred to as the "Tupelo Tongue" of the Coffee Sandstone, and the rock unit ranges through the early and middle Campanian Age. The position of this skull within the rock unit was not sufficiently documented to attempt a more precise date than early to middle Campanian. The streamside locality in which it was found is not otherwise known for extensive occurrences of Late Cretaceous vertebrates, but nearby sites in the Coffee Sand have a well-known shark-tooth assemblage (Case 1991) and a diverse bony fish fauna (Nolf and Dockery 1990) known from their otoliths (tiny carbonate ear ossicles, which serve in maintaining balance).

The Tupelo, Mississippi, *Deinosuchus* skull specimen seems to be an unusual, fortuitous occurrence. It is notable that the skull is densely permineralized by apatite, like typical Atlantic and eastern Gulf occurrences, and unlike the carbonate preservation in the chalks of western Alabama. However, unlike typical Atlantic and eastern Gulf occurrences, this skull specimen is clearly not water-worn, and it did not form in a transgressive lag deposit. Because it was found in approximately the same stratum as shark-tooth accumulations, we may hypothesize that the crocodylian died during marine transgression over a sediment-starved marine shelf. Apparently, as the vertebrate tooth concentration was being transported from offshore, the crocodylian died and was preserved in place by the abundant apatite (phosphatic) mineral dissolved in the waters. This is a rare sort of preservation in the eastern United States.

It is difficult to make a good postdiction (i.e., a prediction of the past) about the living habitat of this particular *Deinosuchus* because it comes with limited site data. The nearby shark species described by Case (1991) are typical of shallow marine waters farther to the east, and Nolf and Dockery (1990) suggested that the 20 species of teleosts (higher bony fish) they identified from otoliths in the Coffee Sand indicated a shallow subtidal, subtropical to tropical marine setting. In addition, at least one hadrosaur (duck-billed dinosaur) fossil is reported from the same general stratum (Daly 1992; Russell 1988), all of which would tend to indicate that a shore was nearby. It seems that the habitat of this westernmost *Deinosuchus* from the eastern side of the continent was similar to that of individuals living far eastward and northward on the Gulf and Atlantic Coastal Plains.

To the north and northwest of the Tupelo site in eastern Mississippi are additional Coastal Plain deposits of Campanian age that could potentially yield *Deinosuchus* remains. In western Tennessee, especially along Coon Creek in McNairy County, is a band of the Demopolis Formation that has a vertebrate fauna (Russell 1988) similar to those found eastward on the Gulf Coastal Plain. Most significantly, the deposit has produced combinations of marine fossils, such as marine turtles and shark teeth, and terrestrial fossils, including a single dinosaur (Bryan et al. 1991), that point toward an estuarine-type setting. Such fossil admixtures are precisely like those found at most *Deinosuchus* sites, and it would not at all be surprising if the crocodylian is discovered in Tennessee (but that has not yet happened). Essentially the same prediction applies to the very limited outcrops of Campanian

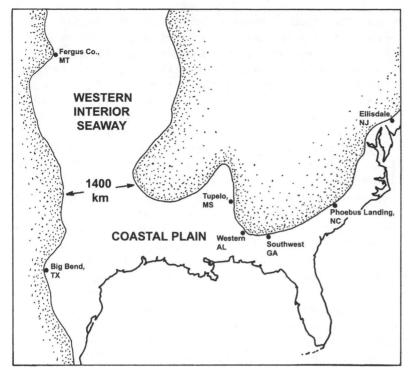

Figure 5.8. Map of the maximum extent of the Western Interior Seaway (WIS) during the early Late Cretaceous (Turonian Age). At such stages of flooding, the narrowest contact between eastern and western sides of the continent was approximately 1400 km. Drawing by D.R.S.

sediment, termed the McNairy Sand Formation, in eastern Missouri. There, both dinosaur fossils (Parris et al. 1988) and marine bony fish teeth (Armstrong-Hall 1999) have been collected, but no documented *Deinosuchus* fossils are known.

These last two Eastern Coastal Plain regions mentioned above—western Tennessee and eastern Missouri—are located at the northern edge of the Late Cretaceous Mississippi Embayment (Fig. 5.8), which is a very long distance from the *Deinosuchus* site in eastern Mississippi. We cannot be sure if the crocodylians came that far north in the embayment. Nevertheless, the overall eastern distribution of *Deinosuchus* is impressive and suggests that they could easily have ranged far up to the northern edge of the embayment. To demonstrate how widespread was *Deinosuchus rugosus* across the eastern continent, consider the following: if one followed a linear track, beginning at the northernmost New Jersey occurrence, down along the Atlantic Coast to South Carolina, westward along the Cretaceous coastal outcrop to Georgia and Alabama, and ending in eastern Mississippi, the total distance traveled would be over 1800 km. As will be shown in the next section, nearly the same distance separates the perimeters of the eastern United States and western United States occurrences of the crocodylians.

Texas

To follow the trail of *Deinosuchus* localities westward from eastern Mississippi, one must make a geographical broad jump all the way to the farthest southwestern corner of Texas. No *Deinosuchus* material of any sort is known from the intervening region, which may reflect either circumstances of stratigraphic preservation or the actual absence of the animals. To begin an argument for a stratigraphic (that is, preservational) cause, I can note that there are no exposed deposits of Campanian Age in western Mississippi and Louisiana. However, there are significant Campanian deposits in Arkansas and eastern Texas, and we must consider why *Deinosuchus* has not been found there.

In Chapter 2, the importance of the Big Bend region of southwest Texas was highlighted, along with the most notorious *Deinosuchus* discoveries. However, the huge area of central and eastern Texas and western Arkansas also includes a very large Late Cretaceous outcrop, containing a broad exposure of Campanian-age beds that are distinct from the Aguja Formation in Big Bend. These represent a paleogeographic region entirely different from Big Bend, which may explain why *Deinosuchus* is absent. For our purposes, it will be instructive to consider the habitats represented in this vast, presumably warm marine area to better understand the preferences of *Deinosuchus,* which lived both east and west of the region we will consider. There is a roughly northeast–southwest line of Upper Cretaceous strata extending from northeastern Mexico up through San Antonio, Austin, and Dallas in central Texas and reaching into northeast Texas and a small area of western Arkansas. These strata formed on the western side of the Gulf of Mexico continental shelf during the Late Cretaceous. However, to the northwest, the sedimentary deposits are contiguous with the southernmost deposits of the Western Interior Seaway (WIS) (Fig. 5.9) because waters of the WIS merged with the northern Gulf. Because this eastern Texas area is transitional between two great, wide marine stratigraphic sequences, it contains the stratotypes (i.e., the reference sections in the type areas) for the best-known North American Cretaceous stratigraphic series: the Coahuilan, Comanchean, and Gulfian. (Such regional stratigraphic usages are a relict of the days when wide-ranging rock correlations were not possible or well understood, and it was necessary to correlate strata within a single sedimentary basin.) Because of their crucial location, the central–east Texas regional stratigraphic names have been frequently used as references outside Texas. Indeed, the terms are still in current common use, even though we now know, for example, that the Gulfian strata are approximately equivalent to the whole Upper Cretaceous Series.

Within the Gulfian Series in the east Texas Basin is the Taylor Group, a term applied to strata of the mid–early through late Campanian Age (Sohl et al. 1991). This set of strata has been assigned to a variety of formational names from one local area to other, but the majority of Tayloran strata are composed of chalk and marl (i.e., impure, clayey, or silty chalk). By their lithology and fossils, we can infer that the exposed Taylor deposits accumulated in deeper waters of the marine shelf within the Gulf of Mexico and the Mississippi Embayment. Some studies have been made of the vertebrate fossils in Taylor strata, including many on the abundant sharks and rays (e.g., Meyer 1974;

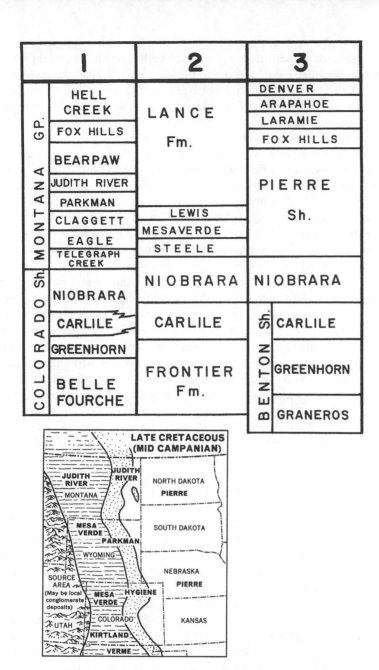

Figure 5.9. Correlation chart and paleolithofacies map of stratigraphic units within the Late Cretaceous Western Interior Seaway (WIS) region. Columns are representative of central Montana (column 1), Wyoming (column 2), and Colorado (column 3). The map below is very generalized and shows the approximate relationship between the marine Pierre Formation to the east and the mixed marine and nonmarine formations westward. Both figures modified and reproduced from Frazier and Schwimmer (1987).

Welton and Farish 1993), the mosasaurs (Echols 1972; Thurmond 1969), and the marine turtles (Zangerl 1953). But very few crocodylians have been found anywhere in the later Cretaceous of the central and western Gulf, and there are no traces of *Deinosuchus*.

Surprisingly, many pre-Tayloran localities in the central Gulf and southern Interior Seaway contain other crocodylian remains, whereas the later Cretaceous deposits, as stated, are nearly devoid of the same. In Chapter 7, I will discuss the ancestry of *Deinosuchus* and examine some of its earlier relatives; but here, in brief, let's consider that many mesoeucrocodylian-grade (i.e., more basal) crocodylians of the family Goniopholidae have been found in Cretaceous deposits of the Aptian through the early Cenomanian Ages, in and around the east Texas Basin and the southern WIS. These ages span the late Early to early Late Cretaceous and are often informally termed the Mid-Cretaceous. They date roughly from 120 to 95 million years in age (Appendix A). The Mesoeucrocodylian occurrences include the Trinity Group in Arkansas, the Glen Rose, Paluxy, and Woodbine Formations in central and northeast Texas (Langston 1974; Lee 1997; McNulty and Slaughter 1969), and the Dakota Sandstone and Kiowa Shale in Kansas (Mehl 1941; Scott 1970; Vaughn 1956). Although these all predate *Deinosuchus* by more than 15 million years, it is interesting to speculate why the more primitive crocodylians are common in a region that the giant eusuchian (advanced) genus apparently could not occupy.

The answer appears to lie in changing regional marine conditions between the Mid- and Late Cretaceous. As discussed in the preceding sections, *Deinosuchus* fossils are predictably rare in open-water deposits, especially marine chalks, because the crocodylians preferred nearshore and estuarine habitats. In general, the regional Mid-Cretaceous deposits containing mesoeucrocodylians formed in environments located relatively high on the continental shelves, whereas the Late Cretaceous, Tayloran deposits formed in deeper, open-sea settings. Indeed, some of the Mid-Cretaceous deposits contain dinosaur bones (Lee 1997), and the Glen Rose Formation is famous for its extensive beds of dinosaur tracks, certainly indicating deposition above sea level! The differences between Mid- and Late Cretaceous conditions would follow the subtle dynamics of sea level fluctuations and basin subsidence. Although it seems odd that not a single stray *Deinosuchus* specimen was transported into deeper water—say, eastward from Big Bend or westward from Mississippi—it is apparent that the deep waters of the central Gulf created a major biogeographic barrier for the big crocodylians. In a following section, we will reexamine the same conundrum for deposits made all along the Interior Seaway, where *Deinosuchus* occurs far to the north in Wyoming and Montana, but not in the intervening marine chalks of Kansas and Oklahoma.

It is evident that the search for *Deinosuchus* requires deposits of Campanian Age with close proximity to marine shores, and these occur on the far side of Texas in the Big Bend region. In Chapter 2, I discussed the discovery of the "*Phobosuchus riograndensis*" specimen in Big Bend and highlighted the nature of the original locality. The discussion to follow will focus briefly on the paleoenvironment of the Aguja Formation in Big Bend and its relationship with other Late Cretaceous regional deposits. During the Late Cretaceous, the southern Big Bend region (approximately centered in the present National Park) lay in a geographically unique position (Fig. 5.8). It was situated simultaneously at the extreme western edge of the Gulf of Mexico Coastal Plain, the extreme south of the Interior Seaway, and at the eastern edge of wedges of sediment coming from the actively rising Cordilleran Mountains in western Mexico. Throughout Campanian time, the marine shoreline migrated east to west, and vice versa, over hundreds of kilometers, rising and falling with sea levels, between continental areas to the west and marine conditions to the east. The result was the Campanian Aguja Formation (see Chapter 2), which contains at least four members (Lehman 1989) that each include some continental, marine, and paralic (transitional) beds. To put this in perspective, during the transgressive stages of the ocean waters, the shoreline migrated west–southwestward and extended Gulf and Interior Sea conditions over Big Bend. During regressive phases, deltaic sands poured out from the western mountains via rivers and prograded (i.e., built outward) over the same region. As the shoreline migrated eastward, conditions changed through paralic to continental. This pattern continued through the Campanian, in an apparently erratic pattern, which left the Aguja Formation with members representing all the intervening conditions.

The Aguja Formation has an amazingly diverse list of species, including terrestrial animals such as amphibians, snakes, lizards, primitive mammals, and dinosaurs, and marine forms ranging from oysters to sharks (Rowe et al. 1992). *Deinosuchus* fossils are found largely in the landward-deposited members of the Aguja, especially the McKinney Springs Tongue in the middle of the Formation. Unfortunately, unlike deposits in the eastern Gulf, the geometry of sedimentation in Big Bend is complex, and it is not easy to determine in a given site precisely which ancient environment is represented by which particular bed. It appears that abundant *Deinosuchus* fossils in the Aguja Formation are found in deposits representing nearshore and estuarine conditions, but this is not easily proven.

Besides an abundance of *Deinosuchus* remains in localities within Big Bend, there is also the interesting proximity of *Deinosuchus* and dinosaurs within the same strata. Remains of many Campanian dinosaur species are common in the Aguja Formation (Lehman 1997), although complete or even partial skeletons are very rare. In Chapter 8, I will consider the most intriguing question of whether or not *Deinosuchus* grew to immense size because of natural selection as a dinosaur predator. Because of the complex intertonguing of marine and nonmarine beds in the Aguja, it is not easy to judge whether *Deinosuchus* ever occurs in continental beds, where the dinosaurs would be expected to be found. We can be quite sure that dinosaurs did not inhabit true marine environments (even though hadrosaurs have often been assumed to be semiaquatic); however, it is apparent that in Big Bend, the *Deinosuchus* populations lived in close proximity to abundant large dinosaurs, probably in the paralic environments representing salt marshes and tidal basins. It is highly suggestive that many very large *Deinosuchus* specimens are found in the Aguja Formation, the implications of which will be discussed in Chapter 8.

Western Interior Seaway

The geography of North America during most of the Late Cretaceous was completely different from modern times, if for no other reason than because a sea existed inside the continental interior. Indeed, the Late Cretaceous continental geography was also different from earlier times of the geological past, when the continent was frequently flooded, but not in the same configuration. For the first time in history, during the Late Cretaceous, an interior sea occupied an elongate basin extending all the way from the Arctic Ocean down to the Gulf of Mexico. The waterway mostly or completely bisected the North American continent. This Western Interior Seaway, as it is formally termed, was not a true ocean in the geological sense that it did not occupy a basin reaching down to the lower crust of the Earth, but it was nevertheless vast (Figs. 1.3, 5.8). At its maximum extent, during high sea level stands, the Seaway was more than 2000 km wide (Hay et al. 1993), extended 6000 km in length, and reached depths estimated to 1.5 km. However, much of it was at more typical Continental Shelf depths of 200 m or less. Most important to our topic, the WIS created a biogeographic barrier that divided North America into two subcontinents. Lehman (1987) termed these the "Asian-American Peninsula" on the west, and "Euramerica" on the east, reflecting evidence that the Cordilleran side of the continent (i.e., the western mountains and coastal regions) had land connections with Asia by way of Alaska, and similarly, the eastern continent's continental shelf contacted Europe via Greenland. When the WIS was at maximum high stages, especially during the early Late Cretaceous (Turonian Age), North America's subcontinents were widely separated into eastern and western units with no known land connections between. At other times during the Late Cretaceous, notably for our purposes the mid-Campanian, there may have been island stepping stones between the two sides of the landmass below the waters.

But *Deinosuchus* and many other marine and terrestrial animals of the Late Cretaceous are found in common on both sides of the Seaway. In the next chapter, we will consider how this situation may have developed and what information it may give about the open-sea abilities of the crocodylians. Here, we will consider where *Deinosuchus* and other crocodylians are distributed in the deposits from the Seaway. As noted previously, Late Cretaceous crocodylians are rare in true marine deposits, most especially in the area occupied by the WIS. But it is not always evident whether a given locality and sedimentary unit in the general region occupied by the Seaway contains "true marine deposits." Sea levels fluctuated cyclically throughout the existence of the Seaway (as in Big Bend and the east), and western mountain-building events episodically sent masses of sediment eastward into the marine basin, developing alluvial plains over what were formerly shallow marine areas. In many circumstances, fossil information and sedimentary deposits must be evaluated very precisely to judge whether a specific rock unit formed in or adjacent to the sea. For example, the occurrence of a crocodylian fossil in the region might represent a freshwater setting during a regressive phase, or a marine setting in a transgressive phase. In Western Interior deposits of obviously marine origins, crocodylians of all types are rare to absent. In paralic and freshwater deposits formed adjacent to the Seaway, freshwater crocodylian fossils are fairly common—and thus one must be careful to consider detailed conditions.

No Campanian-age marine deposits contain *Deinosuchus* or any other known marine crocodylian fossils in the Seaway basin north of Big Bend in Texas and south of eastern Wyoming. Geographically, this area includes north–central Texas, small areas of the Oklahoma panhandle and eastern New Mexico, eastern Colorado, western Kansas, and western Nebraska. The absence of crocodylians is a real anomaly for several reasons. First, the same general region, especially western Kansas and parts of adjacent Nebraska, includes Campanian-age deposits full of vertebrate remains. Second, conditions in the WIS paralleled those of the eastern Gulf of Mexico, where *Deinosuchus* is amply represented. And third, *Deinosuchus* occurs to the south of the Seaway in Texas, to the east of the Seaway, in the eastern Gulf of Mexico, and, as will be discussed, well to the northwest of the Seaway. But it is not found in the main deposits of the WIS.

The most famous and fossil-rich deposits of the Seaway are the upper chalks of the Niobrara Formation, which are of earliest Campanian age and just barely overlap in time with the oldest *Deinosuchus* fossils in the eastern United States. The main area of Niobrara deposition was in the southern–central region of the Seaway, especially in Kansas, Nebraska, and Colorado. In some areas, especially western Kansas, the Niobrara chalk deposits are considered to be *konservat-Lagerstätten,* a paleontological term meaning, literally, holding areas of preservation. In the Niobrara, such extraordinary preservations include sharks with articulated jaw cartilages and vertebrae, sharks with their gut contents present (Druckenmiller et al. 1993), and fossils of huge, complete *Xiphactinus* fish with other large, complete fish in their bellies. The Niobrara Formation in the region also contains a myriad of flying reptiles, diving birds (hesperornithids), plesiosaurs, mosasaurs, and even a few dinosaurs among its vertebrate fossils. Even though it was discussed here that crocodylians are not typical inhabitants of the marine conditions that form chalk, still, one would expect a few to be preserved among such riches of fossil preservation if they lived anywhere nearby. But none has been found.

The Pierre Formation is a mostly dark shale marine unit that overlies and partly intertongues with the Niobrara Formation and spans the remainder of the Campanian in many of the same WIS areas. The areas containing the Pierre Shale include much of the same central interior basin that contained the Niobrara Sea, and the Pierre is also mapped northward into South Dakota, easternmost Montana, and Manitoba. It too has an abundant vertebrate fauna in Kansas and Nebraska (Russell 1988), with remains of innumerable fish, mosasaurs, plesiosaurs, marine birds, and a few dinosaurs. But again, in deposits from the Pierre Sea, there are no crocodylians.

The Niobrara and Pierre Seas occupied the eastern side of the Western Interior Basin (Fig. 5.9). Farther to the west was the front edge of the Cordilleran fold and thrust belt, a newly forming series of volcanic and thrust-faulted highlands running roughly north–south down through eastern British Columbia, westernmost Montana, eastern Idaho, central Utah, and western Arizona. Between the eastern marine trough and the western mountain uplifts were a variety of piedmont and coastal plain environments grading down to the WIS. These mountain-to-marine transitional areas varied greatly in width and location at various times of the Late Cretaceous, depending on the level of the sea and the activity of mountain building. The result is a profusion of sedimentary formations and rock unit names to the west of the Pierre and Niobrara Formations (Fig. 5.9).

One of these formations is the Judith River Formation in north–central Montana and southern Alberta (where it is called the Dinosaur Park Formation), which represents a variety of alluvial (river) and nearshore marine settings accumulated during a 5-million-year span of the middle to late Campanian. The Judith River Formation in Fergus County, central Montana, is the source of the type specimen of *Deinosuchus hatcheri* (Holland 1909), as discussed in Chapter 2. This is also the only likely marine crocodylian occurrence in all of the WIS deposits, aside from anecdotal (i.e., unpublished) reports of *Deinosuchus* teeth and osteoderms in other sites in central Montana and adjacent areas of Wyoming (discussed below).

The Judith River/Dinosaur Park Formation is especially known for its superabundant and diverse dinosaur assemblages in Dinosaur Provincial Park in southern Alberta. In these beds, the dinosaurs are usually preserved in stream channels and overbank deposits (i.e., swamps formed behind overtopped levees). Although these nonmarine dinosaur beds of the Judith River/Dinosaur Park are the most famous and have received most of the intensive collecting efforts (Brinkman 1990), the formation also contains large exposures of vertebrate-rich, nearshore marine beds. These nearshore deposits in the Judith River have many aspects that compare favorably with nearshore deposits in the Aguja Formation in Big Bend, Texas, including the presence of many characteristically brackish-water vertebrates. Such forms include certain sharks (e.g., *Scapanorhynchus,* the goblin sharks), many rays and sawfish, gars and other bony fish, and some turtle species, all commonly found in Campanian-age Gulf of Mexico Coastal Plain deposits (Fiorello 1989; Lehman 1997). However, even nearshore Judith River assemblages still contain many types of nonmarine vertebrates, showing that the depositional settings were located close to shore. For example, Judith River marine fossil sites include dinosaur bone fragments (which were obviously transported; Horner 1989), snakes, lizards, nonmarine crocodylians, and small mammals.

The same notoriety of Judith River dinosaur fossils has led to the common perception that it is a largely nonmarine formation, and as a consequence, the type *Deinosuchus hatcheri* specimen from the Judith River has been commonly considered a nonmarine occurrence. However, because nearshore deposits are present in the unit, the origin of the original *Deinosuchus* site is enigmatic. This is especially true because no other fossils were described in association with the type specimen, and no additional *Deinosuchus* fossils have been formally described and studied from the same site. Similarly, anecdotal reports of isolated *Deinosuchus* teeth and osteoderms in "Judith River" and equivalent deposits (e.g., the Two Medicine Formation in Montana and the Parkman Formation in Wyoming) in the northern Great Plains are fairly frequent, but none has been formally described. The distinctive morphology of both teeth and osteoderms of *Deinosuchus* suggests to me that these informal reports may very well be correct, and it is reasonable to assume that *Deinosuchus* was present along the nearshore areas of both central–western Montana and Wyoming. However, the scarcity of associated remains also suggests the crocodylians were not nearly as common there as in either west Texas or along the entire eastern U.S. Coastal Plain.

To my reasoning, the fact of *Deinosuchus* occurring in ambiguously continental to nearshore deposits in both west Texas and Montana is not coincidental, especially because the intervening, fully marine deposits of the WIS are devoid of the crocodylians. My interpretation is that the huge *Deinosuchus* of the western continent, which occur slightly later than their eastern relatives, were already preadapted to estuarine and nearshore environments. These habitat preferences were then easily conformed to the range of coastal plain and nearshore environments represented by both the Judith River and Aguja Formations. As will be discussed in detail in Chapter 8, the eastern *Deinosuchus* populations were probably engaged in part-time dinosaur predation and would therefore readily take to opportunistically feeding on the more abundant dinosaurs of the American west. The slightly younger populations of western *Deinosuchus* found an ideal ecological niche along the west side of the WIS, where their giant size would naturally select for predation on dinosaurs dwelling in lower river deltas and backswamps, as well on those that were shore-dwelling. The presence of *Deinosuchus* in the coastal plain deposits of both Texas and Montana suggests that their absence in intervening open marine Campanian deposits is not likely a preservational phenomenon.

On the subject of meaningful absences, the absence of any crocodylian fossils in the deeper marine chalks and black shales of the Campanian seaway also suggests that true oceangoing crocodylians were not present at that time. We can be sure it was not a climate phenomenon that results in a lack of crocodylians north of Texas in the Seaway. This is evident both because *Deinosuchus* occurs relatively far north in Montana and because many nonmarine crocodylians are found in contemporary deposits spread from Wyoming to as far north as Alberta. Indeed, one of the earliest alligatoroids, *Albertochampsa*, perhaps distantly related to *Deinosuchus* (Chapter 7), was present in nonmarine beds of the Campanian Belly River Formation in Alberta (Erickson 1972). Also present, and apparently common in freshwater habitats throughout the American west during the Campanian (and surviving into the Tertiary Period), were many species of the crocodylian genus *Leidyosuchus,* along with some more primitive mesoeucrocodylians (Goniopholidae). These and other Late Cretaceous crocodylian taxa will be discussed in Chapter 7, in the context of examining the ancestry and relationships of *Deinosuchus*. The significant point here is that it is evident crocodylians, as a group, were climatically well adapted to the general region of the WIS (Markwick 1998a, b), up to at least the latitude of central Canada.

6. How Many *Deinosuchus* Species Existed?

Were Eastern and Western *Deinosuchus* the Same?

The gigantic *Deinosuchus* specimens in Texas and Montana started the mystique of the giant Cretaceous crocodylian. But as we have seen on the basis of the number of localities and specimens, smaller *Deinosuchus* were probably much more common and widespread across the Atlantic and eastern Gulf coast than were the larger forms in the west. The next theme to address is the relationship between the eastern and western forms and to consider whether or not they were the same species. This question then engenders a natural follow-up series of questions: If they were the same species, why was there such a size difference? If they were not the same species, how different were they? And, overriding these questions: How could members of the same genus evolve on both sides of the vast Western Interior Seaway (WIS), regardless of how many species existed? Did the ancestors from one side swim across more than 1000 km of ocean water? (They would have to travel as gravid females, in pairs, or in larger groups in order to procreate on the other side.) We will first consider the best information available about the species relationships in *Deinosuchus*, followed by the mystery of the WIS crossing.

At the time of this writing, new specimens of *Deinosuchus* from the Big Bend region are being recovered from the field and in laboratory preparation under the direction of Dr. Wann Langston Jr. at the University of Texas, Austin. Dr. Langston has told me (personal communication) that at least one of the better new specimens equals the American Museum of Natural History (AMNH) "*Phobosuchus*" *riograndensis* holotype in size and is much more complete. Perhaps these specimens will reveal new morphologies to fill in gaps in information already available; the discussion to follow can only work with present information and may be modified as new individuals are discovered or elucidated. I take pains to point out this possibility because some very important aspects of our present knowledge of *Deinosuchus* morphology have been known for fewer than three years, and I assume that similar amounts of new information remain to be discovered in as short a time.

To evaluate the relationships among the *Deinosuchus* fossils from various sites, we should start with the diagnostic characteristics of the genus and see how much these details may vary among areas. This will convey a sense of the limits within the generic concept. The reader will recall (or see Chapters 1, 2, and 7) that the genus *Deinosuchus* is diagnosed technically from among other eusuchian crocodylians by the combination of its large size; thick, lumpy, deeply pitted osteoderms; and ruggedly enameled teeth, especially the low-crowned posteriors. In trying to apply this diagnosis to all deinosuchids, we are immediately confounded in the Northern Interior (Montana and Wyoming) by the absence of any reported tooth specimens. Now, as I noted concerning the new Big Bend collecting, it is entirely likely that *Deinosuchus* teeth and skull materials will be found in the Northern Interior. But for the present, lacking comparative teeth and having no information at all about the skull and jaws from what is the genotype reference area of *Deinosuchus,* we draw a blank on those characters in *D. hatcheri.*

Because of the limited fossils and body parts from the Northern Interior, it is unclear whether the same species of *Deinosuchus* was present there and in Texas, but my sense is that it is highly probable the Montana and Texas *Deinosuchus* were one and the same. I base the logic of this assumption on two related premises: first, it is improbable that two giant crocodylians of the same genus would evolve at the same geological time in contiguous marine areas. Evolution does not tend to favor diversification of species within a single genus in a single area, except where there are different environmental niches to fill. For example, in modern Australia, *Crocodylus johnsoni,* a smaller species, inhabits upstream, freshwater inland sites, whereas the large "saltwater crocodile" *C. porosus* is chiefly found in larger, lower rivers and especially in marine estuaries. Where *C. porosus* is present, it seems to inhibit success of *C. johnsoni.* But after extensive hunting of the saltwater crocodiles, where *C. porosus* has been effectively removed by human activity, *C. johnsoni* tends to take over the habitat (Ross 1989). By use of such observations on modern species, it seems improbable that two comparably large deinosuchid species could evolve and coexist in similar habitats along connected shorelines in west Texas and Montana–Wyoming.

The second argument for a conspecific (i.e., same species) relationship among western deinosuchids is related to the first. Given that *Deinosuchus* must have somehow made it across the WIS (discussed in the next section), then a migration along the western shore of the Seaway would have been a relatively simple matter. In the modern world, for example, we see the wide geographic spread of many crocodylian species. Examples are the "mugger" crocodile, *Crocodylus palustris,* spread across the entire Indian subcontinent, and the American crocodile *C. acutus,* which, before human interference, was common from Florida, across the Caribbean, to northern South America. Even more widespread are the Nile crocodile, *C. niloticus,* which ranges across all sub-Saharan Africa, and the living champion among wide-ranging crocodylians, the saltwater Pacific crocodile *C. porosus,* which has a modern range from Southeast Asia to Australia, including most of the intervening islands of the South Pacific. All of these living species have considerably wider geographic ranges than would be necessary to spread *Deinosuchus* from Big Bend to central Montana. Thus, there are perfectly good modern analogs to argue that it is reasonable to assume only one *Deinosuchus* was present in the American west.

The relationship between *Deinosuchus rugosus* in the eastern United States and *D. riograndensis* in the west becomes the crux of our species analysis. When I began the study of *D. rugosus* in the southeastern United States, it was under the assumption that these were a different species from the bigger western monsters (although I did assume they were closely related). Like most workers, I was impressed by the extraordinarily massive size of the Texas and Montana specimens. I also considered the very low crowned, rounded posterior teeth of eastern *D. rugosus,* with their heavy enamel, to be a distinct feature of the eastern species. It was not clear from described specimens in Big Bend whether the same type of rear teeth was found there, and whether the teeth were as ruggedly enameled in proportion to their huge size.

The best information on the relationship of animals from either side of the continent comes from comparing specimens of similar sizes, when this can be managed. By minimizing the effects of allometry (i.e., proportional changes of various body parts with changing size), one may observe whether or not there are real similarities in size-related characters. One of the first insights I received into the cross-continental *Deinosuchus* relationships came from comparing some of the larger osteoderms from the east with the smaller ones available from Texas (Fig. 6.1). I observe that the osteoderms associated with eastern *D. rugosus* teeth and bones are generally more regular and thinner, and they tend to have better-defined keels than we find in specimens from both Big Bend and Wyoming–Montana; this same fact was noted by Erickson and Brochu (1999). Indeed, the original *D. hatcheri* osteoderms (Holland 1909) from Montana both essentially lack keels and are incredibly irregular and lumpy.

Brochu (1999) suggested that these differences among geographically separate osteoderms might be explained by the results of inadvertently comparing 'derms from different body regions—because we cannot be sure where a given 'derm was originally located, having no completely articulated specimens for reference. It may also be argued that the generally larger size of the western *Deinosuchus* specimens is responsible for the lumpier, more irregular osteoderms. This is anatomically reasonable, given that the 'derms are a functional part of the dorsal musculature (see Chapter 3). Recognizing that they help in body suspension, one would expect positive allometry between overall animal size and the proportional massiveness of the osteoderms. Because most eastern animals were smaller than those of the west, their 'derms

might well be more gracile, as we observe. The presence of keels on some eastern osteoderms may also be related to their slenderness: that is, at smaller sizes, the keels would be evident, whereas in the larger versions, the keels become obscured by the bony overgrowth necessary to accommodate muscle stresses. Resolution of the relationships of eastern and western osteoderm morphology became clear when I compared similar-sized specimens (Fig. 6.1). With size-scaling effects minimized, it is obvious that at least some eastern osteoderms are the same as some Big Bend specimens and that assumptions of size affecting the morphology are probably correct. Thus, we may conclude that of the three diagnostic generic characters, the osteoderms show no significant regional variations aside from size-related effects.

Turning to the distinctive dental characteristics of *Deinosuchus*, we may find two major aspects of comparison: tooth counts and individual tooth morphologies. Of these, characterizing the morphology of the individual teeth is part of the formal generic diagnosis, but it is also standard procedure in species-level crocodylian taxonomy to tabulate the number of teeth in various mouth regions (premaxilla, maxilla, and dentary). It has been possible only in recent years to analyze the tooth counts of eastern *Deinosuchus* because we did not have sufficiently preserved specimens available before the mid-1990s. Indeed, most of the available information on *D. rugosus* tooth counts comes from two specimens in my laboratory, combined with the Tupelo, Mississippi, skull specimen (see Chapter 5). However, details of the shapes and structure of individual teeth come from literally hundreds of specimens.

One of the most significant, and startling, comparisons in tooth counts as well as overall size in *Deinosuchus* took place on a visit I made to the vertebrate paleontology laboratory at the University of Texas in Austin. There, Dr. Langston had the AMNH *D. riograndensis* type specimens on loan. Dr. Langston was attempting a more realistic

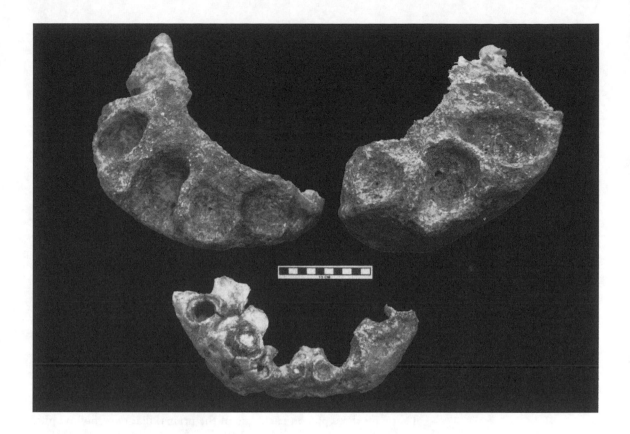

reconstruction of the Big Bend specimen, and I was visiting the lab at his invitation specifically to learn more about the relationship of the *Deinosuchus* materials from either side of the continent. I had brought with me the premaxilla and right half mandible from the Alabama *D. rugosus* specimen (Fig. 5.6) to compare it directly with Big Bend material, including the AMNH type. Dr. Langston placed the AMNH premaxilla pieces in a sandbox, oriented as close to anatomically correct position as possible given the poor state of their preservation and original preparation. He placed the Alabama specimen's premaxilla in front of the gigantic AMNH types, as shown in Figure 6.2. The size comparison was amazing! The Texas premaxillae dwarfed the Alabama material to such a degree it seemed incomprehensible that both could represent adult individuals of the same species.

Because the sense of overall mass of an object is the result of three dimensions (height, width, and depth), mass tends to approximate a cubed function of any one of these three measurements. That is, the ~8.0-m Alabama specimen has a premaxilla that is about 28 cm wide laterally (i.e., across the snout), whereas the AMNH Big Bend specimen was approximately 37 cm wide as reconstructed. But it was also proportionately that much greater in bone thickness and dorsoventral height, producing an overall mass that is 2½ times larger. The effect of observing the size difference first-hand was extraordinarily impressive,

Figure 6.2. Comparison of Deinosuchus *premaxillae from Big Bend (above) and Alabama (below). Notwithstanding the extreme size differences, the positions of alveoli and occlusion notches are basically the same. Note that the preservation of the Big Bend specimen is very poor and that the midline region is missing. Big Bend specimen courtesy of Wann Langston Jr.*

but when analyzed rationally, it is predictable. On close examination, and despite the impressive size difference, the two premaxillary specimens showed all the same tooth positions and characteristics, at least as far as they are preserved. Figure 6.2 shows that they both have four teeth on each side, with two smaller teeth near the midline and two much larger teeth laterally. The difference in sizes of the central and lateral teeth are apparently greater in the smaller Alabama premaxilla, but we cannot be too sure that the AMNH specimen accurately reflects the original sizes of alveoli (tooth sockets) because of its history of poor preservation and preparation. One very interesting and famous set of features of the AMNH premaxilla is the large fenestrations on the anterodorsal surface (see Figs. 2.5, 2.6, 2.8). As I discussed in Chapter 2, it is not clear whether these appeared as a result of incorrect preparation of the difficult material or if they were real anatomical features. And, unfortunately, the Alabama specimen (and all other known *Deinosuchus* premaxillae, to my knowledge) is simply missing this area, and we cannot shed new light on the nature or reality of these holes in the snout. From what remains of the Alabama premaxillary specimen, it is apparent that the bone becomes very thin in the region and could easily be lost in preparation.

Another noteworthy set of features that are clearly evident on both the large and smaller premaxillae are the size and locations of occlusion pits between the teeth. These pits are adaptations to receive the tips of lower jaw (dentary) teeth when the jaws are closed (i.e., occluded). There are three pits on each side of the premaxilla: two shallow pits, one each located between the first and second, and second and third teeth, and a much bigger, better-defined pit located between the third and fourth teeth. All of these occlusion pits are situated slightly behind the upper premaxillary teeth and thereby show us that the upper anterior teeth protrude slightly out beyond the lowers. We can be confident of these interpretations because tooth size and placement in crocodylians tend to follow consistent patterns and because the purpose of the occlusion pits is quite obvious from modern species. In all crocodylians, of necessity, the lower teeth are slightly offset relative to the upper jaw so that they may pass by each other when the jaws close. It is clear that in *Deinosuchus*, the first and second dentary teeth, which were smaller and pointed, fitted into the small pits between the first–second and second–third teeth in the premaxilla. They were offset about one-half space rearward relative to the upper teeth for this positioning to occur. The large third dentary tooth occluded into the large pit between the third and fourth teeth in the premaxilla, and the very large fourth dentary tooth slid into the distinctive external notch between the premaxilla and maxilla, just behind the large fourth premaxillary tooth (Fig. 6.3). All this detail may be discerned on the premaxilla and the anterior mandible (i.e., lower jaw), which is well known from several *Deinosuchus* specimens, both eastern and western. It is also evident from the posterior skull specimens collected in the eastern United States (Fig. 6.4) that *Deinosuchus rugosus* had occlusion pits located all along the maxilla (i.e., the majority of the upper jaw), which were located medially (toward the skull's midline) relative to the upper teeth. This

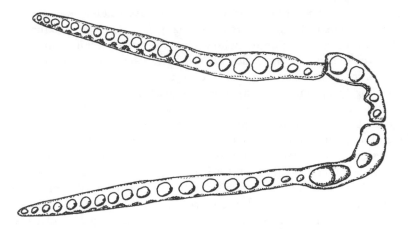

Figure 6.3. Illustration of tooth occlusion in the anterior skull of D. rugosus *showing the offset of teeth 1–4 and the positions of occlusion pits. The upper half represents the upper jaw (pre-maxilla and maxilla), and the lower half of the drawing is the mandible. Drawing by D.R.S.*

indicates that when the jaws were shut, all of the lower teeth behind the fourth had their tips contained within the upper jaws. It is also evident that the upper jaws largely overhung the lower jaws, as in modern *Alligator*, except in the area around the front corner of the mouth where the fourth dentary and third and fourth premaxillary teeth were located. There are grooves present on the outer surface of the dentary, around the front corner, in positions corresponding to the large third and fourth premaxillary teeth; the grooves show that these teeth protruded when the jaws closed. We can therefore interpret that when *Deinosuchus rugosus* had fully closed jaws, the third and fourth premaxillary and fourth dentary teeth protruded externally, whereas all the other teeth were enclosed within the jaws.

Figure 6.4. Posterior maxillary tooth row of the Deinosuchus *skull from the Jones Dam Site. Shallow occlusion pits for the dentary teeth are evident inside the upper tooth row, with a pit offset one-half space from each maxillary tooth.*

These details are not particularly distinctive when viewed among all crocodylians because there are so many variations in the nature of tooth occlusions within the whole group. But the combination of an externally exposed fourth dentary tooth, together with internal tooth occlusion in the remaining skull, seems to be unique to *Deinosuchus*. Among living and fossil crocodylians, occlusion of the lower teeth within the upper jaw is most characteristic of the derived alligatoroids (e.g., *Brachychampsa,* caimans, *Alligator*), whose broad upper jaws tend to slightly overhang the lowers, enclosing all the lower tooth tips. This condition contrasts with the more ancestral character evident in modern *Crocodylus* (i.e., true crocodiles), as well as many Mesozoic and Tertiary crocodylians, where nearly all of the teeth either protrude or are externally visible when the jaws are shut (producing the famous toothy grin of true crocodiles). In addition, nearly all crocodylians, aside from more derived alligators, have the ancestral feature of the external notch between the maxilla and premaxilla to receive the large fourth dentary tooth. *Deinosuchus,* it seems, had both the derived (i.e., advanced) internal occlusion feature of higher alligatoroids (on the basis of the presence of lots of occlusion pits), combined with the plesiomorphous (i.e., primitive) crocodylian feature of an external notch for the fourth dentary tooth. From the evidence of the premaxillae shown in Figure 6.2, both sets of characters are evident in eastern and western *Deinosuchus,* and this seems to be another unifying feature between the populations. At the same time, these combined primitive crocodylian and advanced alligatoroid tooth patterns and positions further distinguish *Deinosuchus* from other Cretaceous crocodylians.

Finally, before leaving the topic of tooth comparisons, we should consider the heavily enameled teeth of *D. rugosus* (which contribute the "rugose" of the species name) and the rounded, blunt posteriors. These teeth are the most common and characteristic *Deinosuchus* fossils found in eastern sites. Do western specimens show the same characteristics that defined the eastern tooth morphology? Because the holotype of *Deinosuchus hatcheri* lacked any associated teeth, the question is moot regarding the generic namesake. Aside from the holotype (i.e., name-bearing) specimen, to date, no teeth attributable to *Deinosuchus* have been formally reported from the Western Interior, nor am I aware of any collected but unreported specimens. Therefore, the tooth morphology of *Deinosuchus* in the northern localities is completely unknown, and to make tooth comparisons, we must consider specimens from Big Bend, Texas.

The holotype of *Deinosuchus (Phobosuchus) riograndensis* from the Aguja Formation did have teeth associated with the mix of bones, some of them intact in the jaw fragments and some loose in the sediment. Unfortunately, none of these teeth from the original "*Phobosuchus*" Big Bend specimen was in a very good state of preservation. One of the best is set in an alveolus (tooth socket) in the largest mandibular fragment (Fig. 6.5). But it is obviously a replacement tooth because the diameter is much smaller than the alveolus and the exposed portion of the cusp is small and clearly unworn. The morphology of this unerupted tooth suggests that when it was fully developed, it would

Figure 6.5. Close-up view of the mandible fragment from the Phobosuchus riograndensis *holotype set, showing a partially erupted replacement tooth in the largest alveolus.*

have a blunt, rounded crown. And from the diameter of the alveolus, it would have been huge: approximately 5 cm in diameter and at least that height. To my knowledge, no one has cut into any of the original *D. riograndensis* teeth to determine whether they have the thick crown enamel of *D. rugosus* teeth. Even if this were done, it is not clear that the mineral preservation of the Aguja Formation would have maintained the internal structure in a sufficient state to make this a definitive test, as discussed below.

There is an interesting conundrum relating to characteristic eastern- versus western-style preservation among Late Cretaceous vertebrate fossils, and it may shed light on the difficulty in making what should be a simple comparison. Most vertebrate fossil occurrences along the eastern United States Coastal Plain tend to be isolated, abraded bones and teeth, largely because they come from marine environments and have been transported via currents and tossed around by sea waves. But at the same time, these bones and teeth tend to be individually well preserved, with hard phosphatic mineralization showing excellent details of the original bone surfaces. Not only are external surface details usually well preserved, but so too are the internal bone structures and the three-dimensional shapes of bones, which are rarely crushed or distorted. In other words, in the eastern Late Cretaceous, we tend to find well-preserved individual bones and teeth, but we consider ourselves fortunate, indeed, to find two or more bones together from the same animal.

In contrast, fossils in Big Bend and many western localities were less aggressively affected by the results of marine water energy, probably because the Interior Sea did not have waves and tides equal to those of the Atlantic Ocean and the Gulf of Mexico. Therefore, many western fossil occurrences consist of associated sets of bones and teeth, and nearly complete specimens are possible to find. However, the majority of western Late Cretaceous fossil sites occur in modern deserts or semiarid regions (e.g., Big Bend), and because of climatic effects on the rocks enclosing the fossils, the mineralization of buried bones and teeth is commonly the product of calcium carbonate salts, rather than salts of calcium phosphate as we find in the east. Carbonate-preserved fossils are not as hard as those preserved in phosphates, and they often tend to be flattened, cracked, and recrystallized by the mineral calcite. The difference in eastern- and western-type preservation is reflected in the colors of typical specimens: most eastern *Deinosuchus* bones and teeth are black or very dark shades of brown or green, whereas western materials are tan to light brown in color (see, for example, Fig. 6.1). Bones tend to preserve adequately well on both sides of the continent, but the finer structures of teeth are usually much better preserved in eastern-type phosphatic conditions. It is because of this preservation difference that we may not be fully able to compare the nature of *Deinosuchus* teeth from Big Bend and the east.

Returning to *Deinosuchus* teeth from Big Bend, aside from the type specimen, other large crocodylian teeth have been collected in the Aguja Formation, and by their size and occurrence in beds containing the diagnostic osteoderms, they are attributable to *Deinosuchus* (Lehman 1997; Rowe et al. 1992). I have seen several of these in the collections of the University of Texas at Austin (Fig. 6.6), and they do indeed show the external striations of the enamel that are characteristic of *D. rugosus*. None of these specimens was cut to reveal the internal structure, but their shape, size, and density indicated that there were no particular reasons to distinguish them from the common teeth of eastern *D. rugosus*. Curiously, among the Big Bend *Deinosuchus* teeth I observed in the University of Texas at Austin collections, none is larger than the common eastern *D. rugosus* teeth, and they are much smaller than the teeth accompanying the "*Phobosuchus*" *riograndensis* type specimen. I am not aware that isolated teeth have been found in Big Bend that would correspond with the size of the "*P.*" *riograndensis* type specimen. It is odd that gigantic western *Deinosuchus* teeth have not been found, aside

Figure 6.6. Comparison of isolated Deinosuchus *teeth from the Aguja Formation in Big Bend, Texas (left two), with* Deinosuchus *teeth from North Carolina (right two). Although the North Carolina teeth are larger, the low-crowned specimens are essentially the same. Specimens courtesy of the Vertebrate Paleontology Laboratory of the University of Texas, Austin.*

from the specimens within the jaws of the skull specimens, and there may be information hidden in that situation.

Given all of the previous discussion, can we answer the question posed by this section's heading? How many *Deinosuchus* species existed? Obviously a definitive answer is not possible because there are limits to our knowledge of the makeup of both eastern and western forms, and these knowledge gaps are not about the same things. But I am more impressed with the similarities between the widespread populations, which have been recently revealed, than with their differences. And I see no differences that cannot be explained by size effects alone. Tentatively, then, I consider *Deinosuchus* to compose a single species, which by publication priority (i.e., on the basis of the oldest designation, in Emmons 1858) should be called *Deinosuchus rugosus*. The next question to investigate is this (Fig. 6.7): how and why did this crocodylian cross the WIS?

Mystery of the Interior Seaway Crossing

One of the most distinctive and pervasive features of North American geography during the Late Cretaceous was the presence of the WIS. I have discussed some aspects of the WIS in several previous sections, but now it will be squarely examined in its relationship to *Deinosuchus*' distributions and species. The Seaway, or at least its precursors, was present in the continental interior for much of Middle Jurassic through Cretaceous time (see Frazier and Schwimmer 1987, chapter 8); but it was not by any means a continuous barrier to cross-continental migra-

Figure 6.7. Deinosuchus *swimming in the Western Interior Sea. The birds are* Ichthyornis, *common to both sides of the Seaway. Painting by D. W. Miller.*

tion during that entire interval. To consider the importance of the WIS to the cross-continental relationships of *Deinosuchus,* we must understand how much of a presence it was through the age of *Deinosuchus,* and this requires some understanding of the nature of the Seaway itself.

To a geologist, a sea is largely a topographic basin (i.e., a depression in the Earth's crust) that happens to be filled with salt water. There were series of topographic basins in the western U.S. continent during the Jurassic and Cretaceous Periods, all located to the east of the rising mountain ranges collectively known as the Cordillera (or the Cordilleran Mountains). The Cordillera trend generally north to south and are roughly parallel to the western continental margin. This parallelism is not coincidental, because plate tectonics activity on the western continental margin is the ultimate cause of the development of the Cordillera. The mountain systems we see today are largely the results of several orogenies (mountain-building episodes) during Jurassic through Early Tertiary times. The interior basins, occupied by the interior seas we are considering, developed as "foreland basins" in response to the rising mountains. The geophysical explanation for the origin of foreland basins is complex and still largely inferential (see Beaumont et al. 1993 for an excellent discussion), but a simplified model suggests that as magma (molten rock) accumulates under rising mountainous terranes, the passive continental crust adjacent to the mountains (part of the craton, or the stable continental nucleus) becomes depleted of magma, and the surface sags, creating basins. Such basins would evolve into elongate troughs, or a single continuous trough, in front of the very long Cordilleran system. They also reached substantial depths below sea level in many areas, in which cases the basins are termed "foredeeps." When the foreland basins extended to either the Arctic or Gulf Oceans, they filled with marine water and formed a seaway.

Several orogenies affected the western continent during the later Mesozoic, notably the Nevadan Orogeny (chiefly in the Late Jurassic), the Sevier Orogeny (Early and Mid-Cretaceous), and the Laramide Orogeny (Late Cretaceous and onward into the Tertiary). Each of these events represents slightly different tectonic phenomena and affected slightly different Cordilleran areas. The events that bear directly on the history of *Deinosuchus* are the two latter events: the Sevier and Laramide Orogenies, which overlapped around late Santonian through middle Campanian times. During these episodes, a substantial interior basin was a continuous presence in the foreland (i.e., on the interior side of the Cordillera) and was undoubtedly filled with salt water. Its presence would have had a major influence on the spread of *Deinosuchus* across the continent, simply as a marine barrier to migration. Because the first appearance of the genus is on the eastern side of the continent (Chapter 5), the following discussion will assume that *Deinosuchus* evolved there and somehow migrated westward.

We cannot be sure when the very first *Deinosuchus* or its direct ancestral crocodylian group existed, but it must have been around or before ~82 Myr (at the beginning of the Campanian Age) because we have specimens of that date in Georgia and Alabama (Chapter 4). In order to evaluate the importance of the WIS to the cross-continental spread of the crocodylians, we need to observe the configuration of the Seaway during the years of their first appearances on both sides; this would be bracketed between a few million years before the beginning of the Campanian in the east, through the first appearance of western specimens in the late Campanian Aguja and Judith River Formations. It turns out that this time interval (~84 to ~77 Myr) coincides with the WIS achieving and remaining at a very wide extent. Kauffman and Caldwell (1993) point out that from latest Albian time (~100 Myr) through the middle Maastrichtian (~70 Myr), high sea levels, combined with continuous subsidence of the foreland basin, ensured that the WIS continuously extended from the Arctic to the Gulf Oceans. The precise width of the Seaway during the entire interval is difficult to model because sea level variations would have caused hundreds of kilometers of shoreline fluctuations. But we must assume the WIS was a substantial water body during the entire existence of *Deinosuchus* and would have required a major open-marine crossing to travel between the eastern and western subcontinents of Late Cretaceous North America (see Fig. 1.3).

Recent discoveries of several characteristically western U.S. fossils from the Santonian and Campanian rocks in the eastern United States (e.g., Schwimmer 1995) show that, somehow, many vertebrate species made it across the Seaway. We are, of course, speaking of vertebrates other than pelagic marine species (i.e., open-ocean dwellers) that would easily and commonly migrate across 1000 km or more of ocean water. Even though *Deinosuchus* was evidently adapted to nearshore marine environments, one would still not presume it was able to make an open-sea crossing of that magnitude. In the modern world, only two species of crocodylians may be considered true saltwater crocodiles: the American crocodile (*Crocodylus acutus*) and the Indo-Pacific crocodile, *Crocodylus porosus* (which is often called "the" saltwater crocodile). Of the two species, only *C. porosus* is commonly observed to venture out into the open ocean. Although there may seem to be similar behaviors and abilities involved in a feeding foray into the open South Pacific by a *C. porosus* and a trip across the WIS by a *D. rugosus,* there are two critical matters to consider in assessing the likelihood: osmoregulation and reproduction.

Modern crocodylians are unable to drink salt water while at the same time successfully maintaining their blood ionic balance at tolerable levels (termed "osmoregulation"). This includes the two saltwater species mentioned above that occupy nearshore marine habitats. This is true despite the fact that most crocodylians have salt glands in their tongues, which facilitate the removal of salt from their blood (Taplin and Grigg 1981). Taplin and Grigg discovered the salt glands and their function in *C. porosus,* and they observed similar structures in *C. acutus* and *C. johnsoni*—that is, in a wide spectrum of living crocodylians. Subsequent research has shown that these glands are adaptations to maintaining osmotic balance during periods of drought (with little freshwater intake) and for long periods of immersion in salt water, where fluids are drawn from the body through the skin. Nevertheless, the salt glands in all species are not sufficiently effective to allow a seagoing crocodylian to simply drink seawater (Ross 1989). And because these glands are present even in freshwater crocodylians, it is evident that they did not evolve as equipment for extensive marine travel. Therefore, unless *Deinosuchus* possessed an unknown mechanism to maintain its blood ionic balance, it would not be able to drink water during a WIS crossing.

The second obstacle to dispersal across the WIS, that of reproduction, is obvious but not to be ignored. Crocodylians have not been observed on open-sea voyages traveling in groups or pairs, nor is this consistent with their behavior. Even if a *Deinosuchus* did manage at times to successfully cross the WIS, that individual would have no effect on distributing the species across the continent simply because it would have been a singleton. For a successful spread of the genus, there would have to be a founder population making its way across the WIS, with enough individuals of both sexes present to assure that mating pairs of the crocodylians could meet and breed. The only alternative scenario is an impregnated female crossing the WIS and laying her eggs on the opposite shore; but this assumes the survival of a viable breeding population from a single individual's offspring.

These difficulties in spreading across the WIS might seem to make such an event very improbable. But nevertheless, *Deinosuchus* and several other nonpelagic (i.e., land bound) vertebrate species were distributed on both sides of the Late Cretaceous Seaway, and therefore they somehow accomplished the feat. Table 6.1 summarizes major parts of the nonpelagic vertebrate fauna that we know existed simultaneously on either side of the Seaway around the time of *Deinosuchus*. Among these animals, ironically, is another crocodylian genus, *Leidyosuchus,* which engenders a detailed but interesting story. This genus (named for the famous 19th-century paleontologist Joseph Leidy) includes probably the most common and widespread of fossil tetrapods in freshwater deposits of Cretaceous age in the western United States. Species attributed to *Leidyosuchus* also range through an unusually long stretch of geological time, from the Campanian Age in the Late Cretaceous well into the Eocene Epoch of the Tertiary. (These crocodylians are thereby remarkable in being among the larger animals surviving through the Cretaceous–Tertiary [K-T] boundary, the infamous mass extinction event!) Recent reassessment of the relationships and taxonomy within *Leidyosuchus* by Christopher Brochu (1997a, 1999) show several interesting results, notably that not all species assigned are really congeneric (i.e., belong in the same genus). Brochu argued that at least some "*Leidyosuchus*" species are actually primitive alligatoroids and should be replaced into another genus (which Brochu 1997a designated as *Borealosuchus*). For our purposes here, the question to address is whether a species of *Leidyosuchus,* or *Borealosuchus,* crossed the Seaway, and we will not delve into details of the proper systematics relating to the clade.

Only recently has a "*Leidyosuchus,*" or similar crocodylian, been recognized in the eastern Late Cretaceous. A small skull and mandible from the early Campanian Mooreville Formation in western Alabama (Fig. 6.8), held in the Auburn University collections, has been the subject of some debate among specialists. It was initially reported (Parris et al. 1997) as a Cretaceous occurrence of the alligatoroid genus *Diplocynodon,* which otherwise is known from the Early Tertiary of Europe. A major reason for this identification is the presence of conjoined third and fourth teeth in the mandible of the Alabama specimen, appearing to come from a single alveolus (Fig. 6.8). This joined dentary tooth pair was in fact the basis of the generic name *Diplocynodon* (*diplo-*, two; *-cynodon*, canine tooth) for the Tertiary genus. However, these conjoined third/fourth dentary teeth are actually a fairly common feature in several crocodylian taxa, especially plesiomorphic (i.e., basal) alligatoroids, including *Deinosuchus* itself, and at least some species assigned to *Leidyosuchus.* I believe this Alabama specimen should be correctly identified as a *Leidyosuchus,* similar to *Leidyosuchus canadensis* (which happens to be the sole species Brochu 1997a considered valid for the genus). My reasoning is that, first, *L. canadensis,* like all species attributed to "*Leidyosuchus*" in the older literature, does show

TABLE 6.1.
Partial list of vertebrates, Santonian and Campanian Ages of the Late Cretaceous,
found in marine deposits on both sides of the Western Interior Seaway.[a]

CATEGORY	TAXON	WESTERN SIDE	EASTERN SIDE
Marine animals	Crocodylia	*Deinosuchus riograndensis*	*Deinosuchus rugosus*
	Turtles		
	Toxochelyidae	*Toxochelys brachyrhinus*	*Toxochelys moorevillensis*
		Ctenochelys (2 spp.)	*Ctenochelys* (2 spp.)
	Protostegidae	*Protostega gigas*	*Protostega dixie*
	Chelospargidae	*Chelospargis advena*	cf. *Chelospargis advena*
	Trionychidae	*Trionyx* sp.	*Trionyx* sp.
	Pelomedusidae	*Bothremys barberi*	*Bothremys barberi*
	Birds	Hesperornithidae (4 spp.)	Hesperornithidae, indet.
Flying animals	Pterosauria	*Pteranodon* (7 species)	*Pteranodon?*
	Birds	*Ichthyornis* (7 species)	*Ichthyornis* sp.
Freshwater aquatic animals	Crocodylia	"*Leidyosuchus*" (4 species)	*Leidyosuchus* sp.
	Osteichthyes	*Belonostomus longirostris*	*Belonostomus* sp.
Terrestrial animals	Ornithischia		
	Hadrosauridae	*Kritosaurus* (*Gryposaurus*) spp.	*Hadrosaurus foulkii*
	Nodosauridae	*Hierosaurus sternbergi*	Nodosauridae indet.
	Theropoda	*Albertosaurus* (2 spp.)	unnamed tyrannosauroid

[a] The list includes nearshore and hemipelagic marine animals, flying animals, and some nonmarine animals that occur on both sides of the Western Interior Seaway, where generally similar forms are in common at either generic or specific levels. If a listing indicates a number of species, it means that several different species are recognized on one or the other side. In some cases, dissimilar genera are listed on either side where the occurrences are especially significant. The list excludes mosasaurs, plesiosaurs (in the larger sense, including pliosaurs and polycotylosaurs), and all marine fishes. It also excludes microvertebrates (e.g., mammals, lizards, and snakes), which are poorly sampled in most marine deposits. Marine turtles are included, although they may be largely pelagic, because they must have land-based nesting grounds. Data come from many sources, most importantly Langston 1960; Russell 1988, 1993; Schwimmer 1995, 1997c; Zangerl 1953; and personal observation.

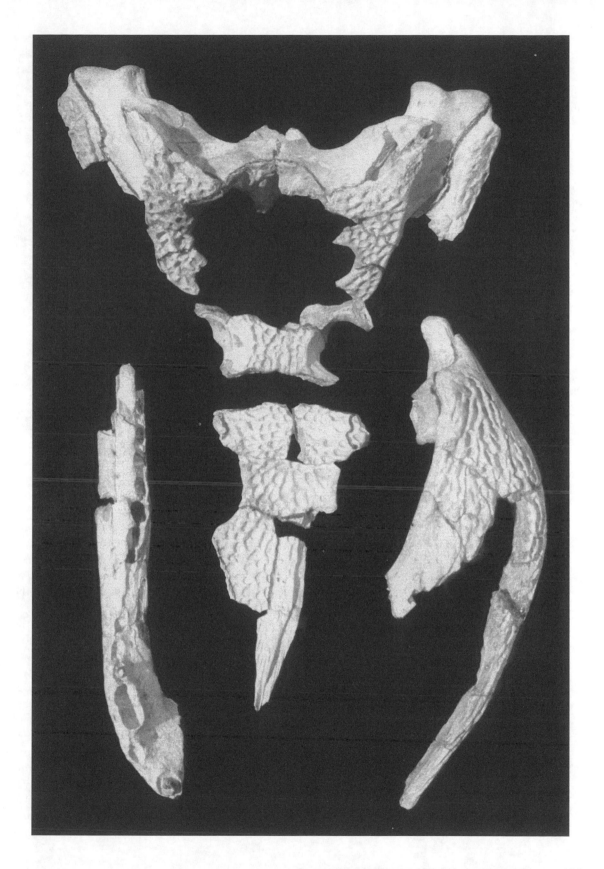

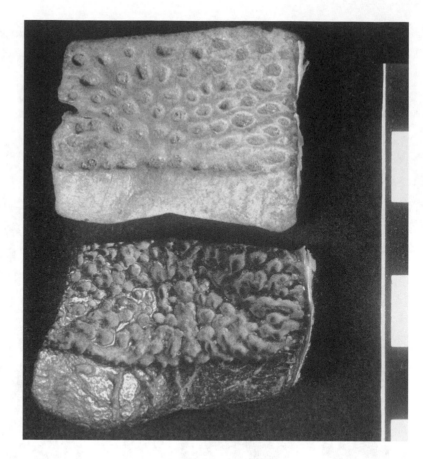

Figure 6.8. (opposite page) Partial skull and mandibles of a small Leidyosuchus *from the Mooreville Formation, western Alabama. The curvature of the rear jaw region (right) is probably due to preservational effects. Note the elongate, doubled alveolus on the anterior jaw unit at lower left. Specimen in the Auburn Museum of Paleontology, Auburn, Alabama. Courtesy of James Dobie.*

Figure 6.9. Leidyosuchus *sp. osteoderms from the Hannahatchee Creek Site, upper Blufftown Formation, Georgia, showing the smooth borders. These are actually fairly small specimens; the scale is in centimeters.*

the enlarged third/fourth mandibular teeth (Mook 1925), which appear to be conjoined. Second, the Alabama specimen has a short mandibular symphysis, with only slight contact by the splenials, which is a distinctive generic feature of *Leidyosuchus* (Mook 1925). Last, *L. canadensis* in the western United States dates to the Campanian Age, like the Alabama specimen.

Reinforcing the evidence for the presence of *Leidyosuchus* in the eastern United States, I have several osteoderms that show the diagnostic characters of the genus (Fig. 6.9). They lack keels (or at best show a vague elevation); they are quite curved and very thin; and some show a smooth border area, lacking pits. This last feature is assumed to occur where two 'derms are shingled and overlap (Erickson 1976), a characteristic of at least the median dorsal series in *Leidyosuchus*. All of these diagnostic features are present in two 'derms from the middle Campanian of western Georgia, and several others show all but the smooth borders. I have also found several smaller crocodylian teeth at the same site, which are not thickly enameled and which may belong to a *Leidyosuchus* species. And I have observed similar osteoderms and teeth in middle Campanian deposits from New Jersey, North Carolina, and Alabama. Indeed, Parris et al. (1997) reported specimens of "*Diplo-*

cynodon" from New Jersey and North Carolina, along with the Alabama specimen, and these too may be additional eastern *Leidyosuchus*. Although "*Leidyosuchus*" therefore seems to have been widespread east of the WIS, the remains are quite rare. This is predictable, because in the west, it is a freshwater genus, and all of the eastern deposits sampled represent marine habitats. It is likely that the individuals we are finding in the east were river-transported specimens (as I assume is the case for much of our well-known but scanty dinosaur record in the east; see Schwimmer 1997b and the discussion to follow).

Another interesting east–west crossover vertebrate group are freshwater fishes of the genus *Belonostomus*. These have wide distribution in the west, like "*Leidyosuchus*" species, and make a rare but unmistakable presence in a single marine formation in the southeast. The first eastern specimen was reported by Whetstone (1978) from Alabama, and the second is in my collections from western Georgia. Both specimens are fragments of the upper jaw, with its diagnostic, elongated rostrum (rather like a swordfish bill in appearance), and both come from the Eutaw Formation, dating between the late Santonian to the base of the Campanian. Neither Whetstone's nor my specimen is sufficiently complete to allow identification to species, but I find no anatomical reasons to separate the Georgia fossil from the common western species *B. longirostris*.

It is difficult to judge the importance of *Belonostomus* in evaluating the likelihood of migrations across the WIS. As apparently freshwater fish, they obviously couldn't swim across the Seaway, which is a factor they share with "*Leidyosuchus*." But a notable difference in the distribution of *Belonostomus* vs. *Leidyosuchus* is their ages of first occurrences: *Leidyosuchus* first occurs in the fossil record around the time *Deinosuchus* appears, at the beginning of the Campanian Age in the Late Cretaceous. In contrast, *Belonostomus* is known in various species from rock that may be as old as the later Jurassic (Estes 1964), and it certainly was present during the Early Cretaceous. Therefore, its presence on both sides of the WIS in the Late Cretaceous is not quite as surprising as that of either *Leidyosuchus* or *Deinosuchus* because there were substantial intervals during the Early Cretaceous when the WIS was not fully present. During times without a full-length Seaway cutting through the continent, land surfaces of the midcontinent were presumably drained by streams containing *Belonostomus* species, and thus they could have spread from west to east.

A complex set of arguments may be made about the cross-continental distributions of several types of dinosaurs of the Late Cretaceous, and these may or may not bear on the paleogeography of *Deinosuchus*. It is obvious that dinosaurs were not marine animals (*Godzilla* notwithstanding), yet we find generally similar groups on both sides of the Seaway. These include some that made first appearances around the same time *Deinosuchus* first appeared, and well after the Seaway became fully (and literally) entrenched. The subject of dinosaurs of the eastern United States has a long and interesting history (see Russell 1997; Schwimmer 1997b; Schwimmer et al. 1993; Weishampel and Young 1996), which is beyond our scope here. One of the salient points, however, is that somehow theropods (i.e., carnivorous dinosaurs) of the tyrannosauroid clade and the herbivorous hadrosaurs (duck-billed dinosaurs) made their first appearances on both sides of the WIS almost simultaneously. Once again, this occurs after the WIS extended fully across the midcontinent from Arctic to Gulf Oceans, presenting us with a conundrum with three plausible solutions. First, both sets of dinosaurs may have evolved separately but in parallel on both sides of the Seaway from older common ancestors. The second alternative is that derived hadrosaurs and tyrannosaurs managed, somehow, to cross the Seaway during the Santonian or slightly older times, perhaps on some sort of land bridges. This alternative implies that we have some aspects of the WIS geometry to better understand. The third possibility is that the dinosaur clades on either side of the continent are not really the same and we are misunderstanding their identities. And, as ever in these multipart analyses, it is possible that one of the three solutions applies to the tyrannosauroids, and a different one applies to the hadrosaurs.

Our knowledge of the eastern dinosaurs is constantly improving as we find better materials, and some of the alternatives mentioned may be eliminated or supported. Recently, for example, it has been claimed (Lamb 1998) that a southeastern "hadrosaur" (*Lophorothon atopus*) described first from Alabama is actually a more primitive dinosaur, an iguanodontid. These were the probable ancestors of the hadrosaurs, and thus Lamb's implication is that *Lophorhothon* was a survivor from an older radiation of ornithischian (herbivorous, "bird-hipped") dinosaurs, which might predate the presence of the WIS. His arguments were based on new fossils showing characters unknown in the original type specimen; and if he is correct, at least one eastern "hadrosaur" is not really such. Rather, it may be descended from a remnant clade of Early Cretaceous age that survived in isolation on the eastern side, whereas its distant relatives evolved into more derived hadrosaurs on the west. However, whether or not *Lophorhothon* is a nonhadrosaur ornithischian, there is no doubt that other eastern hadrosaurs are correctly identified as such. Indeed, the namesake of the family, *Hadrosaurus foulkii,* came from New Jersey (see Weishampel and Young 1996). In addition, I have seen unmistakable hadrosaur fossils, including mandibles with the characteristic multiple rows of teeth, from several southeastern sites. So to understand the biogeography of the Late Cretaceous, we must still confront the spread of hadrosaurs from east to west or vice versa, across the Seaway.

A similar argument may be made for the tyrannosauroids. There are fossils of smaller tyrannosaurs (i.e., smaller than *Tyrannosaurus,* but still large theropods!) in Campanian deposits of Alabama and Georgia. These closely resemble some of the genera in the west such as *Albertosaurus* and *Daspletosaurus* (Holtz 1994; Schwimmer et al. 1993). A well-preserved specimen from the upper Campanian of western Alabama (Fig. 6.10), with about half of the skeleton preserved, is presently being described for the first time and named as a new genus (with myself as a coauthor). Because this study is currently in peer review, it would be inappropriate to report the new name or detailed conclusions, but I may state that cladistic analysis shows this tyrannosaur to be more

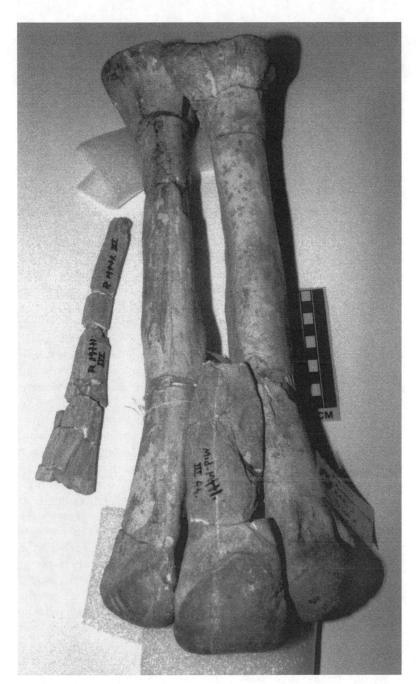

Figure 6.10. Left metatarsus of the unnamed tyrannosauroid theropod from the Demopolis Formation, Montgomery, Alabama. This species is contemporary with Deinosuchus *in the southeast and may represent a new genus. Scale in centimeters; total scale length, 10 cm. Specimen courtesy of Susan Henson, McWane Center, Birmingham, Alabama.*

primitive than those in the west and may well have evolved in place in the southeast from a pre-WIS ancestry. We also know that fossils of large theropods are present in the single Early Cretaceous deposit in the eastern United States from which there is good information, and this may point to the ancestry of these tyrannosaurs. To explain the significance of this information, all later Mesozoic deposits exposed at the surface in the eastern United States are marine sediments from the Late Cretaceous, except for a few outcrops of the Early Cretaceous Arundel Clay in Maryland. This is an apparently mixed marine–nonmarine coastal deposit, and it contains a modest vertebrate fossil assemblage. It also contains the only unambiguous eastern Cretaceous sauropod remains (i.e., from long-necked huge herbivorous dinosaurs), and some arguable teeth from the only eastern representative (and a very early one at that) of the ceratopsian (horned) dinosaurs (Chinnery et al. 1998). Most important here, the Arundel Clay contains teeth from an iguanodontid, and many teeth, phalanges, and other fragments come from larger theropods (Kranz 1989; Lipka 1998). The presence of these latter groups in Maryland during the Early Cretaceous makes it plausible that their descendants evolved independently on the eastern side of the continent. However, one must be cautious. For example, we have no information suggesting that these Early Cretaceous theropods were indeed tyrannosaurs rather than survivors of the carnosaurs, an older theropod clade. If they were carnosaurs, their occurrence would have no bearing on the presence of eastern tyrannosaurs.

One final set of east–west vertebrate relationships is worth noting, and then we will consider a new tack on the bigger question of *Deinosuchus* migrations across the WIS. An obvious difference in Late Cretaceous faunas on either side of the Seaway is the rarity of birds and pterosaurs (flying reptiles) in common between eastern and western fossil deposits. Pterosaur remains have been reported in eastern deposits from only four general ages and localities: the lower Maastrichtian in New Jersey (Gallagher 1984), the lower Campanian in Delaware (Baird and Galton 1981), the middle Santonian in Georgia (Schwimmer et al. 1985), and the lower Campanian in Alabama (Lawson 1975). In all these cases, the remains are no more than three bones of an individual, and two of the occurrences are single bones. For example, in Schwimmer et al. (1985), we reported discovering two bones from the wing finger of a single ornithocheirid pterosaur (Fig. 6.11). This was tentatively identified as a *Pteranodon,* the famous pterosaur with the crest on the head. All of the Santonian and Campanian eastern pterosaur finds are claimed to be ornithocheirids, which were the largest pterosaurs of their times. Following the model of modern seabirds (such as the albatross), big pterosaurs would be the most likely to make a long, cross-WIS flight. What is surprising is that *Pteranodon* and *Nyctosaurus* (another ornithocheirid genus), by far the most common pterosaurs on the western side of the Seaway, have left only scanty traces in the east. Where are the flocks of eastern pterosaurs, if they could indeed fly over the WIS?

An even more inexplicably rare, unique occurrence in the east is a single bone fragment in my collections from Georgia (Fig. 6.12), which is the sole known fossil from a large western vertebrate family. It provides insight into how obscure the occurrence of an animal population can be. Among the abundant and characteristic Late Cretaceous fauna on the western side of the WIS are diving birds in the hesperornithid family. These were large seabirds (pelican- and heron-sized), which are noteworthy for their complete loss of the wings and for their long,

Figure 6.11. Single wing-finger bone (probably a second phalanx) from a pterosaur similar to Pteranodon, *middle to late Santonian, Eutaw Formation, Chattahoochee County, Georgia. Scale in centimeters; total length of specimen approximately 18.0 cm.*

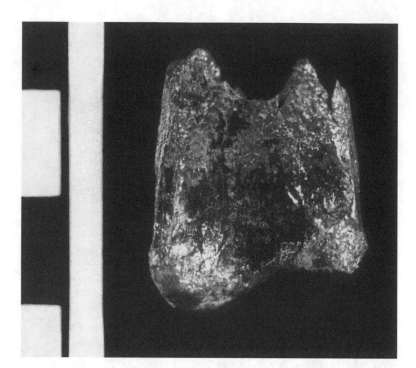

Figure 6.12. Water-worn fragment of the distal tibiotarsus (anklebone) from a small hesperornithiform bird of size comparable to Prohesperornis. Total preserved length 2.5 cm.

toothed jaws; the former was a derived feature, the latter a primitive feature. The namesake genus, *Hesperornis*, is extraordinarily common in the Niobrara Formation of Kansas and in the more northerly WIS deposits of Santonian and Campanian age. Most important to our discussion, and somewhat surprisingly, no specimen of hesperornithid bird has been formally described from any site in the eastern continent (Russell 1988). Because these were marine birds, and because other birds of the same age were clearly able to spread across the Seaway (e.g., the genus *Ichthyornis*), it has always seemed odd to me that the big diving birds were restricted to the west. Yet the single fragment in my collection, a small piece of the tibiotarsus (lower long bone of the leg), is identical to the same bone in one of the smaller hesperornithids. This single bone fragment represents one animal, but that one animal probably represents a population and shows us that we often receive only a glimpse of the true fossil record.

The previous discussion ought to demonstrate that we do not yet have enough information to explain with confidence how *Deinosuchus* and other nonpelagic animal populations crossed the Western Interior Seaway. One hypothesis should be dismissed before continuing further, which is the idea that individual animals frequently took on the notion to venture forth across the uncharted Seaway. This concept was featured in a recent popular television program about dinosaurs and other Mesozoic life. In the TV program, a lone ornithocheirid pterosaur was driven by inchoate instinct to make a unique flight from South America to Spain, to return to its ancestral home. Although a large pterosaur might be able to make such a flight, it seems a poor explanation for dispersal of a nonflying animal group. When ancient species are observed to have spread across a long geographic range, the most parsimonious explanation is that the range was extended gradually, generation after generation, and that we are seeing a phenomenon compressed in the lens of geological time. Our time resolution is rarely less than 100,000 years when we look back to the Mesozoic, and it should not be forgotten that many thousands of vertebrate generations are represented by that number of years. Therefore, applying this logic to our main subject, the idea that lone *Deinosuchus* crossings of the WIS are responsible for spreading the crocodylians across the continent seems to me a last-choice argument.

Finally, arriving at a set of ideas that may answer the question at hand, we come to a reasonable and inclusive explanation: the presence of volcanic islands extending all or partly across the southern reaches of the Seaway. The precursors of this notion are more than a decade old, but only now are data arriving that allow a synthesis that makes sense.

In geological perspective, the WIS was never a true ocean because it occupied a trough floored with continental crust (rather than basaltic, oceanic crust). The detailed nature of the crust under former WIS areas is difficult to know because today, it lies under thick Late Cretaceous deposits resulting from the presence of the Seaway itself (yielding some of the fossils we are discussing). These deposits are typically hundreds to thousands of meters thick, and although they have been frequently drilled for oil and seismically surveyed, such techniques do not give the same grade of detail that can be achieved by studying surface exposures. Fortunately, some deep rock units in the WIS region tend to form structural edifices that are readily observed in seismic profiles, and they may even outcrop (i.e., reach surface exposures). From these we can derive insights into Late Cretaceous volcanism in the WIS basin.

Among the first discussions of possible volcanic events in the Seaway was that of Byerly (1991), who summarized what was known at the time about igneous activity during the Late Cretaceous in the Gulf of Mexico. This included reports of volcanic rocks in an east–west pattern extending from southwestern Arkansas, through Louisiana and Oklahoma, and approximately 240 km into east Texas. The line demarcated by these volcanic rocks runs along the northern margin of the Gulf Coastal Plain and might suggest an ancient volcanic archipelago. However, regarding those rocks, Byerly reported a K-Ar isotopic date of 98 ± 2 Myr, obtained from a single sample along this line of volcanics, which would be too old to directly pertain to the spread of *Deinosuchus*. Nevertheless, the presence of an early volcanic archipelago across the northern Gulf shows that such activity was occurring within the WIS basin. In addition, Byerly (1991) discussed the presence of several additional volcanic sites and regions in the Gulf, including the Jackson (Mississippi) Dome.

The Jackson Dome is a well-defined subsurface structure that happens to lie beneath an economically important reef limestone known as the Jackson Gas Rock (Dockery 1997; Dockery and Marble 1998). The relationship of the Jackson Dome and Gas Rock to our discussion is twofold: first, the Jackson Dome was another volcanic island structure in the Gulf, but it is younger than the line of volcanoes discussed by Byerly. It apparently dates to the age of the Eutaw Formation in Mississippi, which means it was active from the Santonian through early Campanian Ages—the time of first appearance of *Deinosuchus*. Secondly, Dockery and Marble (1998) observed that the limestone of the Jackson Gas Rock represents deposits of a fringing reef that formed around the Jackson Dome (when it was a surface volcano), which may have persisted on the sea surface long after the volcanic core subsided below sea level. This is the pattern of modern volcanic atolls in the Pacific, and we observe the importance of such islands in the distribution of animals through the modern ocean. Although the Jackson Dome is only a single site, it provides a well-studied inference that volcanic atolls were forming on the eastern edge of the WIS, just at the time of the appearance of *Deinosuchus*.

Further extrapolation on the theme of volcanic island activity in the WIS is evident in the suggestion by Cox (1997) that the "Bermuda Hot Spot" may be traced from Kansas to Mississippi. "Hot spots" (or mantle plumes) are localized upwellings of especially hot mantle rock, which are generally held to be the cause of so-called aseismic ridges in the Pacific Ocean (such as the Hawaiian Islands and Emperor Seamount chains). Cox claimed that the ages of volcanics produced by the Bermuda Hot Spot range from as old as 101 Myr in the west (Kansas) to 84 Myr in the east (Mississippi). He reasoned that as the North American continent spread westward by plate movement away from the mid-Atlantic divergence zone, the continental crust rode over the persistent Bermuda Hot Spot, leaving behind a series of volcanic areas. These intracontinental volcanics would have decreasing ages eastward, just as lines of descending-age volcanoes form on the ocean floor of the modern Pacific basin. This same hot spot now lies far to the east of the continental margin, and formed the dome that underlies Bermuda. The interesting aspect of this proposal is that volcanic islands produced by the Bermuda Hot Spot, as with the Jackson Dome, would tend to acquire fringing reefs that would persist long after the volcanoes subsided. Therefore, even though Cox's proposal has the volcanic islands as dating too old for migration of *Deinosuchus* across the Seaway, reefs forming around the islands would probably have persisted well into the Campanian and could have provided stepping stones across the WIS.

A further extrapolation of this island-across-the-WIS theme is Lamb's (1997) proposal that a string of volcanic islands formed an actual, intermittent land bridge at intervals during the Late Cretaceous. And, notably for this discussion, he claimed that the Cenomanian (see Appendix A) and Campanian were likely times for exposure of this land bridge. Lamb's focus was the relationship of such a land bridge to the east–west dispersion of dinosaurs; however, he also claimed that it would provide nesting sites for marine birds and sea turtles (see Table 6.1) within the Seaway. As will be discussed in Chapter 8, turtle nesting grounds could very well have served as feeding stops for *Deinosuchus*.

How shall we put together all the better arguments and come up with the best hypothesis for the cross-continental spread of *Deino-suchus*? The various volcanic island scenarios for a Seaway migration route offer realistic explanations for some of the animal dispersions that we see in the Late Cretaceous, especially the marine species. And because we are primarily interested in the WIS crossing by *Deino-suchus,* reef-bounded, volcanic island-hopping might seem to be the ideal explanation. However, the cross-WIS occurrences of so many vertebrates during the Santonian and Campanian ages suggests to me that the correct explanation for the spread of *Deinosuchus* should also include the joint presence of many other vertebrates we have considered, and there the island idea falls down a bit. Although volcanic islands chains and their reefs might well explain the cross-Seaway migration of marine and flying animals (e.g., sea turtles, seabirds, pterosaurs, and *Deinosuchus*), it is more difficult to envision their relationship to the migration of purely terrestrial or freshwater animals (e.g., *Leidyosuchus, Belonostomus,* and hadrosaur dinosaurs). In the modern world, only a few volcanic archipelagos, such as Japan and Indonesia, have contiguous land surface exposures spanning more than 1000 km, which would be required to span the WIS. Further, although a string of volcanic islands on a map may look like a land connection, for nonmarine animals, the passages and straits between the individual islands would be major obstacles to migration.

Therefore, I do not fully accept the hypothesis that *Deinosuchus* migrated from the eastern to the western United States across a volcanic archipelago, although it might well be true. If future research strengthens the evidence for the presence of very large volcanic island chains, then the hypothesis will be more convincing. But some of the nonmarine life may simply have wandered over or swum through streams on a dry continental interior before the Seaway was fully established in the Late Cretaceous, and we may simply not recognize their earlier presence. This prospect might conceivably include the ancestors of *Deino-suchus* (which we have not yet discovered), which may have evolved the more derived characteristics we attribute to *Deinosuchus* (e.g., osteoderms, size, and teeth) in two parallel populations on either side of the Seaway.

Why Are They of Different Sizes?

In the first section of this chapter, I presented the evidence relating to the number of *Deinosuchus* species that existed, and in that discussion, I concluded that the weight of evidence indicates *Deinosuchus rugosus* is really the same species as *D. riograndensis* and *D. hatcheri.* If that presumption is correct, it both denies the alternative hypothesis presented above (parallel evolution) and forces us to address the striking difference in the sizes of the various *Deinosuchus* populations. As discussed in Chapter 3, *D. rugosus* individuals in the east seem to have averaged 7.0–8.0 m in total body lengths, whereas many *D. riograndensis* (and the paltry *D. hatcheri* remains) were from animals at least 10.0 m long, with some reaching 11.0 m or longer. There is an evident size difference in the populations on either side of the continent, and that difference is too great to be due to chance. Unlike the question of cross-WIS migration, this question of size discrepancy has a ready answer.

Before considering the plausible answer, let me bring up and dispose of two additional explanations that come to mind. The first of these can be dismissed out of hand: that the eastern population is composed of smaller females, whereas the western population is of larger males. Granted, modern crocodylians generally show considerable dimorphic (i.e., sex-based) size difference, and males are larger. Woodward et al. (1995) also showed that the sexes in *Alligator mississippiensis* have different growth rates. It is also true that male crocodylians of several species are aggressive in defending their territories, and we can infer that females of extinct species might have avoided contact with males except at mating times. However, even given all the preceding points, the notion that the sexes of *Deinosuchus* were living on separate shores of a vast ocean is absurd. The distance of separation for a nonpelagic species would have been too great, and swimming such distances for mating would engender all the problems of migrating across the WIS that we have considered. Worse, this marathon swimming would be just at the time males would need all their strength for intraspecific battles and the act of mating! This is not a successful evolutionary strategy and must be rejected.

Another possible explanation we can bring up and then dismiss is the perception that insufficient sampling of the eastern population accounts for our failure to find the larger individuals. From my own collecting and museum work, I can argue against this idea. When we survey the totality of known collections containing eastern *Deinosuchus* fossils, there is a consistent maximum size of specimens that appear, and that includes extrapolation from the enormous number of teeth that have been collected. In my collections from western Georgia and Alabama alone, there are at least 100 *Deinosuchus* teeth, only one of which may come from an animal approaching the size of larger western specimens (Fig. 6.13). Furthermore, this assemblage of teeth probably represents a dozen or more individual animals, when we consider the marine conditions under which they accumulated (see Chapter 5).

One may argue that the sample of eastern *Deinosuchus* specimens is somehow biased against larger individuals because of ecological selection. For example, if it were true that larger individuals lived farther upshore, then we would expect to find fewer of them in the eastern marine deposits. But I reject this possible collecting bias because the paleoenvironmental conditions affecting the eastern *Deinosuchus* population were basically similar to those affecting *Deinosuchus* in the west, especially as we interpret the Aguja Formation in Big Bend, Texas. Because we find many larger deinosuchids in the Aguja, we should find them in the analogous habitats in the east, if they were around. There may be some differences between these strata too subtle to be detected in ancient rock, but that does not seem sufficient reason to explain the dramatic size discrepancies in their crocodylian fossils.

Figure 6.13. Largest-diameter Deinosuchus *posterior tooth known from the eastern United States. This specimen is very fragmentary and has been roughly restored to its original base diameter of ~5.0 cm. This specimen is comparable to the diameter of the large alveolus in the AMNH "Phobosuchus" specimen from Big Bend (Fig. 6.5), suggesting that a few very large individuals were present in the east.*

Having discounted sex separation and bad population sampling, how, then, do I explain the very significant size difference between the deinosuchids from either side of the Seaway? By a combination of direct evidence and inference, I believe it is based on the size of the typical prey of the two separated populations. This topic will be covered in detail in Chapter 8, but to summarize the ideas here, it seems that the eastern population was feeding opportunistically on a wide variety of vertebrates living in the Atlantic and Gulf coast estuaries and bays. Most of these prey species were animals of less than a half ton in size, with side-necked sea turtles as especially common prey. In contrast, adult western *Deinosuchus* fed largely on shore-dwelling dinosaurs, especially hadrosaurs, which included adult animals weighing several tons. There are many shades to these contentions, and it is apparent that eastern *Deinosuchus* occasionally fed on smaller dinosaurs (as I described in the opening chapter of this book; and see Chapter 8). But the size difference in the crocodylian populations reflects the frequency with which the western individuals were eating very big prey—and for which they evolved larger sizes. Increased predator size is a logical response to the more abundant food resource presented by large dinosaurs, and it would help the crocodylians bring down and dismember larger animals. Thus it would be an evolutionary advantage, which may explain why the western *Deinosuchus* population persisted slightly later in geological time than did the eastern.

7. A Genealogy of *Deinosuchus*

Archosaur Relatives and Some Complex Taxonomy

In the Introduction to this book, I gave a brief explanation of the term "Crocodylia" in order to explain why the more common "crocodilian" was not the best to apply to *Deinosuchus rugosus*. From that discussion (or turn back, if you are a preface skipper), the reader will recall that I stated *Deinosuchus* was not a true crocodile, nor an alligator, nor a gharial (which together are the complement of living crocodylian families); rather, *Deinosuchus* was an early offshoot from the ancestry of modern alligators. Therefore, it was an alligatoroid in current terminology, but not an alligator in the strict sense, and certainly not a true "crocodile" in any sense. In this chapter, I will examine some of the interesting animals that were ancestral to crocodylians such as *Deinosuchus,* as well as some groups that were close to the roots of the alligatoroids. In doing so, we will be able to more logically consider the place of *Deinosuchus* within the Crocodylia and understand what is properly mean by the term "Crocodylia" itself.

The common impression of a "crocodile" or an "alligator" is related to the synapomorphies (i.e., shared derived features) of the modern species, many of which are quite unlike the basal features of their ancestors. When we envision a crocodylian belonging to any living group, we expect to see an elongated skull with an especially long, flat rostrum (snout), lots of simple, conical teeth, and elevated eyes set high on the skull and far to the rear, all the better to peer from the water while the animal is submerged. If we are quite knowledgeable about the anatomy of modern crocodylians (collectively termed the Eusuchia), we may be aware that they have a secondary bony palate, unique among reptiles, which allows breathing while the mouth is submerged or clamped around prey. For the postcranial regions of our typical crocodylian, we envision a long body, both low and wide; a large, heavy, laterally flattened tail adapted for sculling through the water; a series of dorsal dermal bony plates and projections; heavy body scales; and relatively short legs. But again, these are characters of higher crocodylian families (in cladistic jargon, termed "crown groups") that are contained within the Eusuchia. It may not surprise students of nature and the fossil record to learn that some early crocodylian ancestors possessed few or none of these traits and that these are not the plesiomorphous (i.e., primitive) characters of the basal crocodylians.

The terminology of higher crocodylian affinities becomes turgid when we include the deep ancestors along with more derived forms as part of the overall concept of Crocodylia. In current taxonomy (Benton and Clark 1988; Brochu 1999; Clark 1996), the term "Crocodylia" strictly refers only to three living families (crocodiles, alligators, and gavials) and their close ancestors. The older term for the order Crocodilia is presently replaced by Crocodyliformes (in the vernacular, crocodyliforms), which includes the living crocodylian families, as well as all extinct Eusuchia plus more primitive extinct groups included in a clade termed the Mesoeucrocodylia (discussion to follow). As Table 7.1 shows, current workers also recognize an even more general group that contains all the crocodyliforms as well as some Triassic animals that were probably the earliest members of the lineage: this largest, most general grouping is termed the Crocodylomorpha (in the vernacular, crocodylomorphs). In this book, I have, and will continue to use, the term "Crocodylia" as the best descriptive category for *Deinosuchus* and all of the similar animals discussed so far. However, I recognize that as a crocodylian, *Deinosuchus* is at the same time a derived crocodyliform and an even more derived crocodylomorph, if we trace back its monophyletic ancestry in referring to its classification. However, in this chapter, for purposes of clarity, I will use the terms Crocodylomorpha, Crocodyliformes, and Crocodylia, largely in the vernacular form, in their stricter senses.

The earliest crocodylomorphs are of early Late Triassic age, dating to approximately 225 Myr. These animals look so little like modern crocodylians that it is somewhat difficult to be sure they are indeed their monophyletic ancestors. In fact, among the list of typically crocodylian features described in previous paragraphs, none is necessarily present in the Triassic ancestors aside from the simple teeth (which are a common basal feature of most reptiles, both primitive and derived). Most of the stereotypical modern crocodylian features are adaptations to aquatic life, and it is evident that the early ancestors of the eusuchians were terrestrial animals. But the Triassic crocodylian ancestors do show some shared derived features (synapomorphies) that are found in all members of later crocodylian clades, chief among which is the possession of a unique ankle structure, termed the "crocodile-normal tarsus." This ankle structure, along with a collection of additional skeletal traits, provides the anatomical basis for recognizing the earliest crocodylomorphs.

To set the stage, we should know that the Crocodylomorpha are a clade within a larger set of reptiles commonly termed the Archosauria

TABLE 7.1.
Taxonomy of *Deinosuchus*, in quasi-Linnean format.[a]

CLASS: REPTILIA

Infraclass: Archosauromorpha
Superorder: Crocodylomorpha
Orders: Crocodyliformes
Suborders: Sphenosuchia
Protosuchia
Mesoeucrocodylia
Subsuborders: Ziphosuchia
Unnamed southern taxa
Neosuchia
Infraorders: Atoposauridae
Goniopholidae
Thallatosuchia
Pholidosauridae
Tethysuchia
Eusuchia
Crocodylia
Gavialoidea (= Longirostrines)
Crocodyloidea
Crocodylidae
Tomistoma
Alligatoroidea
Deinosuchus
Leidyosuchus
Diplocynodontidae
Alligatoridae

[a] Modern phylogenetic taxonomy recognizes such a large number of lower divisions (i.e., below the level that approximates the older concept of the Infraorder) that traditional Linnean classifation schemes have become quite unworkable and perhaps are now irrelevant. Here, I attempt a compromise between the sense of order conveyed by higher Linnean systematics while accepting the variable levels of relationships evident in modern cladistic theory. Levels of crocodyliform relationships below Infraorders are not given hierarchical terms; rather, they are placed in the best approximations of natural groupings. Data leading to this compilation include Benton and Clark (1988), Brochu (1999), and Clark (1996).

(which translates as "ruling reptiles"). Archosaurs are diagnosed as those diapsids (i.e., reptiles with two large openings, or temporal fenestrae, in the posterior skull) possessing a large cavity in the skull anterior to the eyes (an antorbital cavity), and a pelvis with ilium, ischium, and pubis forming roughly 120° angles around the acetabulum (Witmer 1997). Archosaurs also have teeth set into sockets (termed the thecodont condition) and tend to have dermal armor or osteoderms on their dorsal surfaces. As with crocodylian taxonomy, modern workers recognize higher, more inclusive systematic levels of archosaurs with the addition of various ancestral groups. Thus one may recognize the Archosauriformes to include several Triassic lineages, and an even higher taxon, the Archosauromorpha (Benton and Clark 1988; Gauthier 1984; Sereno 1991). This latter concept delimits an enormously diverse group of animals that may be defined to include forms as diverse as birds, dinosaurs, pterosaurs, crocodylomorphs, and a host of varied and sometimes bizarre Triassic forms without common names (e.g., rhynchosaurs, aetosaurs, poposaurs, rauisuchids, prestosuchids, and phytosaurs). Although from a strict cladistic viewpoint it is logical to recognize that all the terminal lineages derived from basal archosaurian ancestry belong within the Archosauromorpha, this is a cumbersome and confusing practice because of their incredible diversity. It is apparent that these creatures all share common ancestry, but if we separate the extremely derived groups from the more basal, the terminology becomes more manageable. If we take that approach, it is evident that dinosaurs and birds make a natural derived group, commonly termed the Dinosauria. Of course, this taxonomic concept also implies that if dinosaurs are "reptiles," then we must consider birds to be reptiles; this conundrum is representative of the debate between conventional and cladistic taxonomic systems and is not likely to be resolved soon. Fortunately, the crocodylians are not so far derived from their ancestors that we would consider separating them from the archosaurs, and so we can avoid some debate here on the arcana of taxonomic theory.

Among the Triassic archosaurs (in the loose sense) are several groups with the specialized ankle found in all crocodyliforms. This ankle structure (Fig. 7.1) is sufficiently unique to suggest to many workers that all animals sharing it have a common ancestry. However, not everyone accepts the significance of the ankle in recognizing lineages of archosaurs (e.g., Dyke 1998). And—unfortunately for simplicity's sake —many archosaurs possessing this same ankle are apparently neither ancestral to crocodylians nor particularly close. These latter, non-Crocodyliformes, seem to be early divergent branches from the common ancestor with the specialized ankle structure, and thus, we may conclude that the "crocodile-normal tarsus" is a necessary but not sufficient means to diagnose the main stem of crocodylians. This ankle morphology is easily recognizable because of its distinctive offset caused by the main hingement running obliquely between the astragalus and calcaneum (Fig. 7.1), with a peglike element of the astragalus plugged into a socket in the calcaneum. The result is a foot that is able to rotate outward and downward around the astragalus–calcaneum articulation as the leg is extended. Alternatively, if the crocodylian leg is held semi-

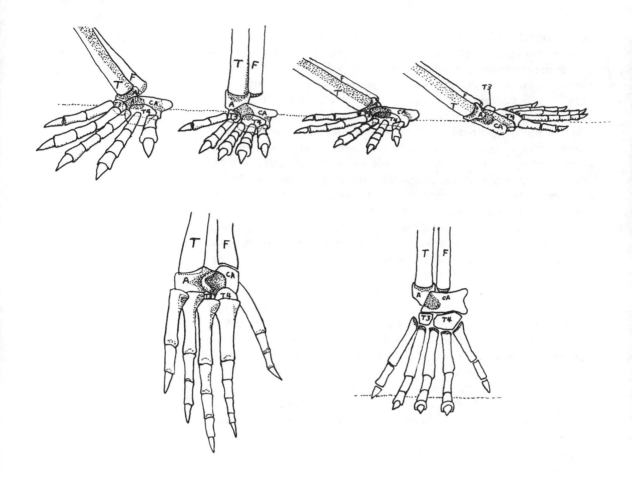

Figure 7.1. The kinetics of the crurotarsal archosaurian ankles. (upper figures) Illustrations of the mobility afforded by the offset structure of the crurotarsal ankle structure, ranging from high-walk to sprawling orientations. (bottom left) "Crocodile-reverse" ankle in which the calcaneum plugs into the astragalus. (bottom right) "Crocodile-normal" ankle in which the astragalus penetrates the calcaneum.
A = astragalus;
CA = calcaneum;
F = fibula;
T = tibia;
T3, T4 = tarsals in sequence.
Drawings by Ron Hirzel.

flexed, the ankle bends forward, which effectively shortens the leg and widens the stance laterally. This ankle mobility gives crocodylians the option of walking in either a short-stride, sprawling posture, with the legs folded around the ankle, or in the longer-stride high walk, where the ankle untwists somewhat, the legs are pulled inward medially, and the effective leg length is increased (Fig. 7.1).

However, there is another variation of this rotary ankle joint among Triassic archosaurs. In this ankle morphology, the calcaneum plugs into the astragalus (Fig. 7.1), which is essentially the reverse of the above, and hence it is commonly termed the "crocodile-reverse tarsus." Cruickshank (1979) and Parrish (1986) argued that both of these ankle structures evolved from a more primitive archosaurian ankle in which there were two pairs of articulations present, suggesting that both ankle structures shared the same ancestry. Both of these offset ankle morphologies have been termed "crurotarsal" because the crus (i.e., the shank of the leg) and the tarsus are involved. Partly on the basis of this synapomorphy, Sereno and Arcucci (1990) and Sereno (1991) have recognized a taxon called the Crurotarsi, which largely contains all the archosaurs except pterosaurs, dinosaurs, birds, and their direct ancestors. Brochu (1997b) commented on the concepts involved in several stem group names for the collection of offset-ankled archosaurs and pointed out that they are all redundant or otherwise problematic in some aspect. My impression from reviewing these concepts, and attempting to summarize them here, tends to echo Brochu's ideas. In my experience, a scientific concept that doesn't hold up well to explanation probably has some fundamental problems in its logic.

Regardless of the relationships among all the early archosaurs, we may tentatively accept that clades bearing the same direction of offset ankle as the crocodylians form a natural group, which is commonly termed either the Crocodylotarsi (Benton and Clark 1988), or, in slightly earlier usage, the Pseudosuchia (Gauthier 1984; Gauthier and Padian 1985). By either name, the clade is defined to include all the archosaurs that are closer to crocodiles than to birds. However, even within this more limited clade, there are still a lot of Triassic-age groups only vaguely related to the Crocodylomorphs. Among the reptile lineages closest to the basal crocodylomorphs were the aetosaurs, the phytosaurs, and three generally similar-appearing families: poposaurs, rauisuchids, and prestosuchids. Again, let me emphasize that none of these clades is likely to have included the monophyletic ancestors of the crocodylomorphs, but rather they probably derived from the same ancestry (Table 7.1). To briefly characterize some of these noncrocodylomorphs, the aetosaurs were quite distinctive animals (Fig. 7.2) and possibly the first archosaurian herbivores. They were of moderate size (1 to 3 m) and low slung, with small heads featuring beaklike or upturned snouts. Their most distinctive feature was heavy dermal armor, often with spiked projections on the lateral margins. In some aetosaurs, the armor formed a dorsal carapace, similar in appearance to an armadillo's. (The dermal armor of the aetosaurs was mentioned in a previous discussion in Chapter 3 in the context of the osteoderms of crocodylians, assuming both are manifestations of the tendency toward developing dermal armor as an archosaurian plesiomorphy.) Aetosaurs were widely distributed in the Upper Triassic of North America, Greenland, South America, North Africa, Europe, and India (Heckert and Lucas 1999) and were probably not ancestral to any post-Triassic group.

The second set of noncrocodylomorphs, the poposaurs, rauisuchids, and prestosuchids, have many similar characteristics and are difficult to classify and separate. Alcober (2000) presented a recent summary of the history of names and classifications assigned to this set of archosaurs, and it is indeed a daunting history. Overall, these archosaurs were large carnivores (some over 9.0 m in length), with long, narrow skulls and sharp, recurved teeth. In many genera the premaxillae were overhanging, producing a snag-tooth appearance. They also had fairly upright four-legged posture, similar to the high walk of crocodylians, which suggests they were agile. On the basis of their size, formidable teeth, and the appearance of mobility, one would assume they were the dominant predators of the late Triassic. Paul (1988) termed some of these large predators "protocrocs," which is not literally true because they were not monophyletic crocodylomorphs; but he correctly pointed out that they coexisted in the late Triassic with the

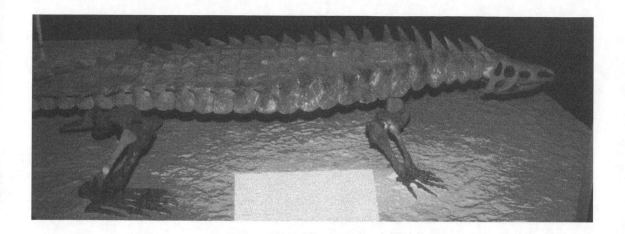

Figure 7.2. Aetosaur Typothorax, *overall length ~2.0 m, showing the elaborate dorsal armor.*

earliest carnivorous dinosaurs, and clearly dominated them. This was especially true in South America, where there coexisted both the oldest dinosaurs and specimens of the large genus *Prestosuchus* (Fig. 7.3), a formidable rauisuchid. Rauisuchids, which featured small dorsal ossicles down the center of the back and tail, are also found in the Middle and Upper Triassic of Europe, Africa, and North America. Also in North America, the carnivorous archosaurian niche was filled by the Middle Triassic *Postosuchus,* a large poposaurid. This animal superficially resembled a carnivorous dinosaur in many skull and limb proportions, with relatively short forearms and long legs; however, the ankle morphology shows it was among the crocodile-normal archosaurs. In fact, poposaurids seem to be the group closest to the basal crocodylomorphs because they possessed a eustachian canal in the skull (part of the ear apparatus), like crocodylians and unlike most other basal archosaurs (see discussion to follow).

The phytosaurs, also called the Parasuchia, the third crocodile-normal clade we are considering, appear externally very much like derived crocodylians and might seem the most reasonable early ancestors. However, at least one significant feature shows they were a derived crown group (i.e., specialized in their own direction) and could not plausibly be the ancestors for crocodylomorphs. Phytosaurs show nearly as many aquatic specializations as most derived crocodylians, including very long rostrums with an extraordinarily large number of sharp teeth (~ 35 to each side). One significant difference in the rostrums of phytosaurs and crocodylians is in the proportions occupied by the various bones: in phytosaurs, nearly half of the rostrum is composed of the premaxilla, which may also contain half of the teeth in the skull. Crocodylia have a much shorter premaxilla, rarely with more than six teeth in each side. In crocodylians, the elongate rostrum is composed largely of the maxillary and nasal bones. Like crocodylians, phytosaurs also reached large sizes, with lengths reaching up to 10 m. Typical phytosaurs, such as *Rutiodon* (Fig. 7.4) of the Late Triassic of North America, had very narrow jaws with slender and pointed teeth, presum-

ably adapted for snatching fish in the manner of modern gavials. Some phytosaurs had wider jaws and serrate teeth, presumably for processing the meat of tetrapod prey (Hungerbühler 2000). But all phytosaurs had elongate, short-limbed bodies, with large, flattened tails, heavy dorsal armor, and a unique breathing adaptation. It is this latter feature that demonstrates unequivocally both their aquatic specialization and their deviation from crocodylian ancestry. Phytosaurs had nostrils at the top of the skull, usually quite elevated and far to the rear, situated on the frontal bones (Fig. 7.4). Assuming that a soft palate separated the nostrils from the mouth, this morphology allowed them to lurk, eat, and breathe while the rest of the body and skull were submerged. Crocodylians have the same abilities; however, they have an entirely different adaptation for this same function. In modern (and fossil eusuchian) crocodylians, the external nostrils are located on the anterior dorsal surface of the skull, and the internal air openings (termed choane) are located far back in the skull (Fig. 7.5), separated from the mouth by a bony palate that serves as a snorkel. More primitive crocodylians have a less complete secondary bony palate (see discussion of the Protosuchia, to follow), but they still show no signs of the adaptations for aquatic breathing that we observe in phytosaurs. The very different biomechanical solutions by phytosaurs and crocodylians to the same breathing needs shows that they are not closely related clades. Nevertheless, phytosaurs were very successful Late Triassic animals and are common in freshwater deposits of both eastern and western North America and of Europe. But like the other crocodile-normal archosaurs discussed so far, they apparently left no post-Triassic descendants.

Figure 7.3. Rauisuchid Prestosuchus, *overall length approximately 3.0 m. (The structure that resembles an eye is a sclerotic ring, which supports the soft tissue of the eye, and is often preserved in archosaurs.)*

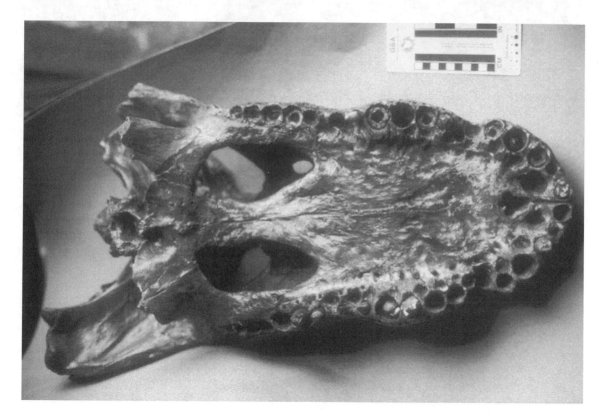

Early Crocodylomorphs

It is not easy to recognize the earliest crocodylomorphs among the wide variety of Triassic archosaurs that are known in the fossil record. It requires that we select anatomical characteristics that are truly diagnostic of the group; but because the early crocodylomorphs hardly resemble the general concept of modern crocodylians, it becomes necessary to consider some fairly arcane skeletal features. Further, these characters have to be independent of the aquatic adaptations that are found in all modern crocodylians, but that may not be basal characters of the larger group. To figure out which are shared derived characters (synapomorphies) of crocodylomorphs, we must examine the features of both derived eusuchian crocodylians and some more primitive but still clearly crocodylomorph Mesozoic groups. It is especially important to find characters that are unique to all these crocodylomorphs, but at the same time distinct from character states of the other archosaurs. When we do all this, it becomes evident that there are many shared derived characters among our sample, especially in the skull. These provide a starting point in the search for basal crocodylomorph features, although we may not expect to find them particularly well developed in the earliest representatives.

The skulls of all derived crocodylians are heavily built, with thick, tightly sutured bones. Crocodylians have nonkinetic skulls; that is, there is no movement possible among the skull bones, such as one finds to a high degree in snakes and lizards. Many extinct archosaurs had some kinesis (motion) evident in the skulls, an inference based on their lightly sutured joints and especially the slender bones surrounding their typically large skull openings (temporal fenestrae). Highly derived crocodylian skulls (i.e., eusuchians) are quite the opposite: strongly buttressed internally in cross section, across the rostrum, with the bony palate forming part of the structural joint. Busbey (1997) referred to this structure of overlapping bones by the engineering term "scarf joints." Another strength-related feature of the typical crocodylian skull is its relatively small fenestrae, especially when compared with other archosaurs; indeed, the eusuchian crocodylians have lost the external openings for the antorbital cavities, which in most archosaurs are exposed through large antorbital fenestrae. Other generally recognizable characteristics of all crocodylians include the unique, squared-off shape of the posterior skull (the skull table), which is quite flat in posterior and dorsal views and is strongly braced and tightly sutured, especially around the braincase. Given the consistency of all these characters, it would seem that a very rugged, rigid skull is an essential feature of at least the more derived crocodylomorphs.

Another diagnostic skull feature, traceable even to the earliest crocodylomorphs, is the shape and orientation of the quadrate bone. This bone forms the upper jaw hinge, and in turn, this quadrate morphology relates to the use of the jaws. Crocodylomorphs have unusually long, strongly muscled jaws, which can open wider than those of typical carnivorous archosaurs. And at the same time, they have the ability to quickly snap the jaws shut. In part, these abilities are due to the large size and position of the quadrate, which extends quite far down ventrally (Fig. 7.6), while also firmly contacting many of the posterior skull bones. By this unique orientation, the crocodylian quadrate effects a strong lever that allows a wide opening of the jaws and provides an extended hinge for the musculature that produces strong snapping jaw closure. The snap of crocodylian jaws is powered by a two-branched, oversized set of pterygoideus muscles that are oriented nearly parallel to the jaws. Contraction of these muscles provides a quick oblique pull on the mandible, in contrast with the slower perpen-

Figure 7.4. (opposite page top) Skull of the fish-eating phytosaur Rutiodon, *showing the external nostrils at the top of the skull, located quite far to the posterior. Skull length ~1.0 m.*

Figure 7.5. (opposite page bottom) Ventral view of a Crocodylus *skull, showing the characteristic feature of the internal choane (nostrils) bounded within the pterygoid bones (located at the rear center of this damaged specimen).*

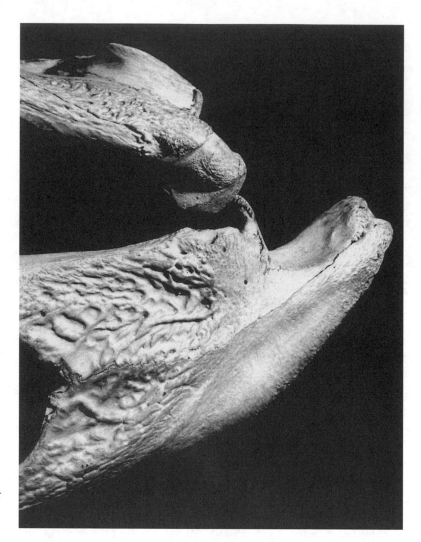

Figure 7.6. Lateral view of the quadrate and articular joint in Alligator. *Note the downward curve of the quadrate and the substantial posterior reach of the retroarticular behind the jaw joint. (This specimen has some of the remaining soft tissue within and around the joint.)*

dicular contraction of the even more powerful adductor muscles (which provide the multi-ton jaw closing power of crocodylians). The posterior pterygoideus in crocodylians wraps around the elongated retroarticular process at the rear of the mandible (Fig. 7.6), adding even more leverage to the jaw snap by means of an extra long pull behind the quadrate hinge. The quadrate must be exceptionally strong to brace all the forces exerted by the oblique tension of the pterygoideus muscle as well as the crushing pressure of the main jaw adductors. An elongate retroarticular process is not present in the earliest crocodylomorphs (the sphenosuchids, discussed below), but we do see the modified quadrate at that stage, and we assume they had at least the rudiments of the snapping jaws of crocodylians.

A final skull feature that seems to unite the crocodylomorphs and to separate them from other archosaurs is the presence of a eustachian canal in the ear system. The eustachian system in all higher tetrapods is an air passage between the middle ear and the pharynx (throat), which derives from a homologous opening in the ancestral amphibian and fish gill systems. Most archosaurs, including the derived dinosaurs, have a cavity representing this air passage, and it is commonly a sizable opening. However, two unrelated groups, mammals and crocodylomorphs, have the eustachian passage modified into a well-defined tube system. Despite this feature seemingly in common, it is very unlikely that the presence of a eustachian tube system unites mammals and crocodylomorphs because their last common ancestor traces back to the Pennsylvanian Period (more than 100 million years before the first mammals or crocodylomorphs appeared). In addition, the eustachian tube of mammals is a simple unbranched unit, whereas that of crocodylians branches in a complex manner as it enters the middle ear (Romer 1956). But the presence of a eustachian tube in some Triassic archosaurs provides a useful character in recognizing early crocodylomorphs.

The oldest and most basal archosaurs that show some of the essential crocodylomorph features are the Sphenosuchia. In current cladistic systematics, these are considered to be the sister group of the Crocodyliformes, and together with them, they comprise the Crocodylomorpha (Table 7.1). Sphenosuchids were long-legged Middle Triassic to Lower Jurassic archosaurs found in nearly all continents; they show no real aquatic modifications. Indeed, the names of some genera indicate their apparent habitats and morphology: e.g., *Terrestrisuchus* ("land crocodile") and *Gracilisuchus* ("graceful crocodile"). However, their skulls did have long, strong quadrates, bent slightly down ventrally, and joined tightly with the adjacent quadratojugal and jugal bones, as in crocodylians. In addition, the antorbital cavity was usually fairly small, and the posterior of the skull was relatively flat. Overall, the skulls of sphenosuchids were less kinetic than those of any other Triassic archosaurs, and thus more like Crocodyliformes. To cinch the relationship, it was observed in an unusually well preserved specimen of the Chinese sphenosuchid *Dibothrosuchus* that the otic (ear) capsule and eustachian canal were essentially the same as in derived crocodylians (Wu and Chatterjee 1993). In the postcranial makeup, there are a few additional clues to their affinities. There are just nine cervical (neck) vertebrae in sphenosuchids, as in all later crocodylomorphs, featuring stout, short ribs from the third cervical vertebra on back. Also, the shoulder girdles of sphenosuchids, in the few specimens where they are preserved, lack clavicles and feature elongate coracoid bones as in later crocodylomorphs. Finally, there appears to be a well-developed series of osteoderms in at least the better-known genera.

However, there are also many noncrocodyliform features in the sphenosuchids. The limbs tend to be long and slender, with the hindlimbs longer than the forelimbs. These limb proportions have suggested to some researchers that sphenosuchids may have walked partially or entirely bipedally—that is, upright on two legs, like dinosaurs. One well-known taxon, *Saltopsuchus,* appears to be the most basal of sphenosuchids and has a generic name that translates from Latin as "jumping crocodile" because of its long-leggedness. The skulls of sphenosuchids, also significantly different from crocodyliforms, lack the aquatic adaptation of a secondary palate; indeed, the well-preserved specimen figured by Wu and Chatterjee (1993) shows no sign at all of its development. From this lack of specialized breathing adaptation and from the general body proportions, it seems evident that sphenosuchids were fully terrestrial animals adapted as swift pursuit predators. Nevertheless, it appears that many crocodylomorph skull characters were already established in the sphenosuchid grade of crocodylomorph evolution. We can speculate that these skull structural details evolved to allow strong biting forces as part of the sphenosuchids' predatory behavior. This adaptation for strong bite pressure in these terrestrial animals shows that it was not a primarily aquatic trait, but it was available for use and modification when crocodyliforms took to the water in the Jurassic.

The evolution of the Crocodyliformes (i.e., crocodylomorphs above the sphenosuchid level, containing the Protosuchia and the Mesoeucrocodylia) is not a simple example of increasingly more advanced animals appearing over time. The more primitive group, the Protosuchia, shows only a few features that are much advanced beyond the sphenosuchid grade, but they overlapped in time with the first of the much more advanced Mesoeucrocodylia. This apparent anachronism is not too surprising when we consider their paleoecology: it is evident that these two early grades of crocodyliforms adapted to different ecological niches. The Protosuchia were small creatures (typically <1.0 m long), with body and limb proportions indicating that they largely continued the older trends as terrestrial predators, like their presumed sphenosuchid ancestors. In contrast, the early mesoeucrocodylians appear in a wide variety of sizes and morphologies (some very large), with most showing aquatic specializations and some being fully adapted to the marine environment. Therefore, these two groups of early crocodyliforms were neither competitors nor cohabitants when they overlapped in time.

Protosuchians are a group of mostly uniform animals found on both sides of North America, as well as in Patagonian South America, southern Africa, and China. They occur in rocks from the Upper Triassic to the Lower Cretaceous (Gow 2000), although most are found in the Lower Jurassic. Protosuchians appear to have changed little through their long time of existence (Fig. 7.7) and therefore must have filled some specialized niche favoring smaller, not quite aquatic crocodyliforms. Buffetaut (1979) portrayed them as amphibious animals, occupying a niche like that of modern river otters. The body proportions were less adapted for running than were those of the sphenosuchians, with shorter limbs and more plantigrade (i.e., flat-footed) posture; however, the limbs of genera such as *Orthosuchus* were still significantly longer than in modern crocodylians. The vertebrae were slightly amphicoelous (i.e., concave at both ends), like those of basal tetrapods and sphenosuchids and unlike the condition in more derived crocodyliforms. The skulls of protosuchids, however, best reveal the advances beyond the sphenosuchid grade. The choane (internal nostrils) opened about midway back in the rostrum, anterior to the palatine bones and quite unlike the higher crocodylian condition. But there was a small

internal bony shelf developed from the maxillary bones, located toward the anterior rostrum, which reinforced the front of the skull and may have helped to resist strong biting forces as in the more derived crocodylians. Also in the skull, the antorbital fenestrae (the openings for the antorbital cavities) were reduced to small vestiges. Another crocodyliform refinement was a deep notch at the junction of the premaxilla and maxilla, into which a large dentary (lower jaw) tooth fit when the jaws were closed, just as we observe in many higher crocodylians.

*Figure 7.7. Representative protosuchians (*Protosuchus *sp.). Drawing by Ron Hirzel.*

Mesoeucrocodylians

It is not clear whether the Protosuchia were an evolutionary stage in the evolution of higher crocodyliforms or just a divergent branch from the lineage that extends to the Mesoeucrocodylians. These latter were termed "Mesosuchians" in traditional taxonomy (e.g., Buffetaut 1982) and assumed to include all of the crocodylomorphs above the Protosuchian grade and below the grade of the Eusuchia. However, cladistic analysis (e.g., Benton and Clark 1988) shows that it is more logical (i.e., nonparaphyletic) to include the Eusuchia as a higher group within the larger clade containing the "Mesosuchia," rather than as a taxon at equal level. The result of this taxonomic rearrangement is to recognize a huge clade of Mesoeucrocodylia, incorporating more than a dozen families of higher crocodyliforms. Given this concept, the Mesoeucrocodylia are much closer to the general concept of "crocodylian," yet they are also a highly diverse group, including many aquatic lineages as well as a few that are terrestrial. *Deinosuchus* is obviously among the Mesoeucrocodylia, as are all other eusuchians.

In the following discussion I will use the term "basal mesoeucrocodylian" to mean lineages less advanced (i.e., less "crocodylian") than the Eusuchia. Basal mesoeucrocodylians are most characteristic of the Jurassic Period and show the initial development of several advanced crocodyliform features. The internal nostrils (choane) are usually located far back in the skull, typically at the front of the pterygoid bones. (However, this condition is even more advanced in the eusuchian grade, where the choane are entirely bounded within the pterygoids.) Basal mesoeucrocodylians also have a partial bony secondary palate developed from the maxillary and palatine bones (but not the complete secondary palate of the eusuchians). A noteworthy primitive feature of lower Mesoeucrocodylia is their retention of amphicoelous dorsal vertebrae (i.e., without a ball-and-socket articulation), characteristic of basal archosaurs as well as the Protosuchia and Sphenosuchia.

In attempting to describe their general nature, we should begin by observing that the mesoeucrocodylian clade is a many-stemmed, phylogenetic bush. There is such a wide diversity of shapes and habitats represented that it is difficult to infer the interrelationships and evolution directions within the overall group, if indeed all the lineages included are truly descended from a common ancestry. In attempting to characterize the Mesoeucrocodylia, it is typically necessary to jump anecdotally from one subgroup to another (e.g., Buffetaut 1979; Carroll 1988; Molnar 1988) simply because the linkages are indeterminable. Ortega et al. (2000) have proposed a basic division between several primitive groups generally restricted to the Southern Hemisphere and some more derived groups with global distribution. There is no unifying name for the southern group (which may not be monophyletic), but it includes the Ziphosuchia (first defined by Ortega et al. 2000) combined with the widespread genus *Araripesuchus*. The second group is termed the Neosuchia (Benton and Clark 1988) and is by itself a very diverse, probably polyphyletic assemblage that includes many relatively advanced lineages, as will be described below. Because the Neosuchia includes the Eusuchia, which in turn includes the Alligatoroidea and *Deinosuchus*, it will be a primary focus in the discussions to follow.

The Southern Hemisphere mesoeucrocodylians above do not include the ancestry of *Deinosuchus*, and I will only briefly consider them. These animals are found in parts of the former Gondwana supercontinent, especially in Africa and South America, and more recently in Madagascar. In fact, the obviously close relationships among the crocodylomorph populations of these continents has been part of the evidence that they remained interconnected into the Mid-Cretaceous, long after the Triassic breakup of Pangea. It is evident that the crocodyliforms in Gondwana remained largely isolated from evolutionary events taking place among their equivalents in the Northern Hemisphere, and that to some degree this isolation continued on into nearly modern times. Among the southern assemblages are many smaller crocodyliforms with short snouts and very large orbits exemplified by the common genus *Araripesuchus*, found in both Africa and South America. The overall body form of *Araripesuchus* suggests they were mostly terrestrial animals and not grossly different from the Protosuchia; but several characteristics clearly show they were more derived toward the higher crocodyliforms, including reduced fenestrae, especially the antorbital and temporal fenestrae located on the high, flat dorsal skull table of the posterior skull (Fig. 7.8). Also, they were among the most primitive crocodylomorphs known with pitted dorsal osteoderms, but they lacked ventral 'derms. These terrestrial crocodyliforms were quite successful in the Early and Mid-Cretaceous and probably filled the small-predator niche during the age of large dinosaurs. It is interesting to note that some *Araripesuchus* species are found in the

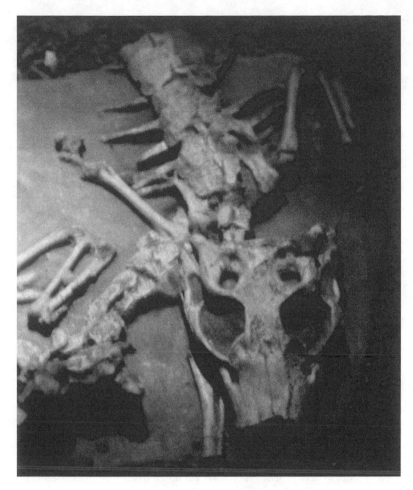

Figure 7.8. Basal mesoeucrocodylian from Gondwana, Araripesuchus patagoniensis, *showing an early stage of the typically crocodyiform flat skull table. Specimen in the Museum of the University of Neuquén, Argentina.*

same Middle Cretaceous beds in Argentina as the carnivorous dinosaur *Giganotosaurus,* which was, by a meter's length, the largest land carnivore that ever existed (as far as we know). In addition, they coexisted with sauropod dinosaurs of lengths over 30 m! Clearly, in that time and place, the role of a giant terrestrial carnivore with even larger prey was preempted by dinosaurs and was not available for a crocodyliform to occupy.

As constituted by Ortega et al. (2000), the Ziphosuchia includes Cretaceous genera such as *Notosuchus* and *Malawisuchus,* which were also small, short-snouted crocodylomorphs, generally similar to *Araripesuchus* but lacking their oversized orbits. The ziphosuchian clade also includes much larger, long and narrow-snouted Gondwanan forms called the Sebecosuchia (formerly called the Sebecidae), which were the last mesoeucrocodylians to survive. The name "Ziphosuchia" refers to the bladelike, often serrate teeth (technically, "ziphodont") found in at least part of the dentition of most members of the clade. In some of the smaller taxa, the complex denticulation of the rear teeth suggests that they were actually herbivorous because the tooth morphology resembles that of both modern herbivorous iguanas as well as some her-bivorous dinosaurs. A bizarre crocodylomorph, recently discovered in Madagascar (Buckley et al. 2000), has fantastically derived ziphodont front and rear teeth that were clove-shaped and multicusped. In addition, the snout is so short that the animal appears to be pug-nosed, and the skull is high and rounded: indeed, it scarcely resembles any concept of a crocodyliform, except for the recognizable osteoderms. This small Madagascene creature was almost certainly an herbivore. A possibly related African genus, *Malawisuchus* (Gomani 1997), also carried atypical tooth morphology to the extreme. It too had an especially short snout, but with complex, differentiated (i.e., various shaped) teeth, superficially resembling those of synapsids (i.e., "mammallike reptiles"). *Malawisuchus* even had mammallike multicusped rear teeth, which, combined with evidence from the jaw articulation, suggests that it evolved a feeding behavior that involved chewing and grinding food as dogs do, or perhaps like the chewing of ruminant herbivorous mammals.

Ziphodonty was well developed in a different direction in the Sebecosuchia, which had a dentition reminiscent of the thin-bladed serrate teeth of theropods (carnivorous dinosaurs). Reinforcing the impression of a theropodlike existence, the skulls of sebecosuchids were laterally flattened and high, with eyes on opposite sides of the skull, like dinosaurs and unlike typical aquatic crocodyliforms. The sebecosuchids were also the last terrestrial crocodylomorphs in the fossil record, and the namesake genus *Sebecus* apparently survived and flourished in isolation in South America, when it was separated from both Africa and the Northern Hemisphere during much of the Tertiary Period. *Sebecus* filled the role of dominant large predator there until the Pleistocene, when North American predators (e.g., big cats) entered South America via the Central American land bridge. When *Sebecus* became extinct, less than 1.0 Myr ago, it was the last remaining meso-eucrocodylian below the neosuchian grade—and at the same time the last remaining crocodylomorph below the eusuchian grade!

Those other Mesoeucrocodylians, the Neosuchia (Table 7.1), are another vast, poorly constrained group, even when separated from the Gondwanan clades we have already discussed. The term "Neosuchia" was coined by Benton and Clark (1988), and the diversity of forms within the taxon is at least equal to the variations among all other crocodylomorphs combined. The Neosuchia includes some fairly basal forms, some extremely modified marine clades, some nearly eusuchian clades, and the Eusuchia themselves. The marine groups are the most unusual and will be briefly considered first. They were the most distantly related to *Deinosuchus,* but they had a number of convergent characteristics, including marine adaptations, their often large sizes, and in at least one taxon, evidence of feeding on marine turtles. During the Jurassic through Early Tertiary Periods, several crocodyliform clades underwent full marine adaptation, with some reaching such an extreme degree that they were fish-tailed and had limbs nearly or truly unable to walk on land. At least two separable, probably unrelated

main lineages of marine mesoeucrocodylians evolved: the thallato-suchians and the dyrosaurs.

All of these marine neosuchian clades share a few characters, espe-cially the tendency to develop highly elongate bodies with short limbs (an obvious streamlining advantage for marine life) and the tendency toward developing very elongate, narrow jaws adapted to feeding on fish by quick snapping of jaws (Fig. 7.9). Thallatosuchians existed only during the Jurassic to early Cretaceous and were extraordinarily spe-cialized marine crocodyliforms. They include two families, the metrio-rhynchids and the teleosaurids, with both showing a host of advanced marine specializations and skulls that usually featured very large open-ings (supratemporal fenestrae) on the dorsal skull table behind the or-bits. Metriorhynchids, in particular, were the most specialized marine crocodylomorphs of all time, evolving fishlike (technically, reversed het-erocercal) tails for propulsion, with the posterior tail vertebrae kinked ventrally in *Geosaurus* and *Metriorhynchus* and a fleshy upper lobe developed to complete the forked tail. They also had front limb flippers with loose, unarticulated bones, like those of ichthyosaurs, and com-plete loss of the dorsal osteoderms (see p. 56 for discussion of osteo-derms related to marine adaptation). Metriorhynchids may have never left the sea, even to lay eggs, and may have borne live offspring by ovoviviparity (internal egg development), like many snakes.

Teleosaurids had less extreme aquatic adaptations than metrio-rhynchids, retaining a typical complement of heavy osteoderms and legs that had fully articulated bones, and they were able to walk on land. The overall form of teleosaurids was distinctive, with hind legs that were much longer than the front legs and extremely long, slender jaws, often with more than a hundred teeth. Some genera had special-ized teeth suggesting predation on animals other than fish, such as the crushing-type teeth in *Machismosaurus* and the multicusped teeth in *Rhytiosodon* (Steel 1973). The anatomical differences between the two different groups of thallatosuchians have been attributed to habitat

Figure 7.9. Drawings of representative thallatosuchians. (top) Geosaur, showing the fishlike tail morphology, loss of osteoderms, and limbs dedicated to fully aquatic propulsion. (bottom) Teleosaur, showing the long hindlimbs, full osteoderms, and the long, slender rostrum for catching fish. Drawings by Ron Hirzel.

preferences (Hua and Buffetaut 1997): it is likely that the metriorhynchids were open-water marine pursuit predators, living a dolphinlike existence, whereas the teleosaurids were postulated to have been shallow-water stealth predators, like modern gar.

Pholidosaurs were a less diverse group of neosuchians that generally resemble the teleosaurids and that survived into the Middle Cretaceous. Like the teleosaurs, they had some less exclusively aquatic features, such as walking limbs and relatively heavy dorsal osteoderms. They also featured very long, narrow rostrums, like teleosaurs, but with the technical differences that the nasal bones contacted the premaxillaries (rather than the maxillaries, as in the teleosaurs), and the supratemporal fenestrae were smaller, of typical neosuchian size. A common feature of several pholidosaur genera was a broadened premaxilla, giving a slightly duck-billed appearance. Most of the pholidosaurs were freshwater aquatic animals, typically found in the Early Cretaceous of Gondwana, and some of these reached enormous size. The genus *Sarcosuchus* (Buffetaut and Taquet 1977), of both Brazil and Niger, was estimated by Buffetaut (1979) to have reached 11.0 m length with a skull length of 1.8 m (thus equaling *Deinosuchus*). More recently, Sereno et al. (2001) estimated the length of *Sarcosuchus imperator* to reach 11.5 m, based on a newly discovered African specimen with a skull length of 1.6 m. They also extrapolated a maximum body weight of 8 tons for the genus, which equals the size of both larger *Deinosuchus* and *Purussaurus* (Chapter 3). These extraordinarily large sizes, in this basal mesoeucrocodylian, show that taxa well below the Eusuchia could become gigantic.

But not all pholidosaurs were restricted to freshwater and the Gondwanan continents. A late-surviving genus, *Teleorhinus* (not to be mistaken for a teleosaur), has wide distribution in early Late Cretaceous marine deposits, including North America and Europe, and it is found as far north as Saskatchewan in the Western Interior Seaway area (Cumbaa and Tokaryk 1993). *Teleorhinus* was a narrow-snouted, 7.0-m neosuchian known in Saskatchewan from a nearly complete specimen and probably was a nearshore marine dweller, like *Deinosuchus*.

The dyrosaurids have an uncertain relationship with the thallatosuchians and pholidosaurs, but it is likely that all these marine neosuchian groups independently evolved their aquatic specializations. The dyrosaurids also include some very large animals, especially *Dyrosaurus*, with skull lengths to 1.8 m, and *Phosphatosaurus* of the Early Tertiary of Africa, a nearshore marine genus that reached 9.0 m (Buffetaut 1979). Like the thallatosuchians, dyrosaurs were long-snouted and long-bodied; however, *Phosphatosaurus* had robust jaws and some blunt, striated teeth (like *Deinosuchus*), suggesting it preyed on marine turtles (Hua and Buffetaut 1997). This feeding preference is one that will be reexamined in the next chapter because it is shared in common with *Deinosuchus*. Dyrosaur specimens occur fairly commonly in the Late Cretaceous deposits of the eastern United States, notably the genus *Hyposaurus*, known from many specimens in New Jersey (Parris 1986) in deposits of the latest Cretaceous (Maastrichtian Age) and Paleocene. In the Late Cretaceous of the eastern United States, dyrosaur vertebrae

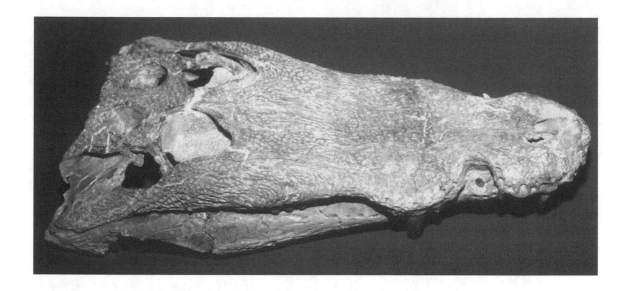

can be easily distinguished from those of the more common eusuchians because they are amphicoelous (eusuchians have procoelous vertebrae; see discussion to follow). To date, no dyrosaur fossils have been found in deposits containing *Deinosuchus,* probably because the latter occurs in the Campanian Age, approximately 8 million years older than the oldest occurrences of *Hyposaurus.*

The nonmarine neosuchians include several less-derived groups that appear to be heading toward the eusuchian grade, as well as the Eusuchia themselves. It is uncertain which among the more basal groups are the direct ancestors of the eusuchians, and it is tempting to look for them among the clade that most closely resembles eusuchians, which would be the Goniopholidae. These were very common in the Late Jurassic to Upper Cretaceous, with species of *Goniopholis* (Fig. 7.10) distributed in every North American marine region that contains Cretaceous marine or aquatic deposits. Goniopholids were mentioned in Chapter 5 (p. 100) as the crocodyliforms most frequently found in the Western Interior Seaway of the Early to Upper Cretaceous—for example, *Dakotasuchus* (Mehl 1941; Vaughn 1956) in the Dakota Formation of Kansas. They are also found in the Upper Cretaceous of New Jersey and North Carolina, and the Early Cretaceous of Arkansas and east Texas (however, some of these reports are based on identification of isolated vertebrae and teeth, which really can't distinguish goniopholids from teleosaurs). In overall skull and body form, goniopholids resembled modern Nile and Indo-Pacific crocodiles. The skulls were broad toward the rear, tapering to fairly narrow snouts, featuring a deep notch for the fourth mandibular tooth and a slightly flaring premaxilla (Fig. 7.10). The body shape and dermal armor were also similar to modern crocodylians. The larger species had skull lengths up to 70 cm and overall lengths extrapolated to 4.0 m. On the basis of their form and occurrences, we assume that their habitats and behavior were similar to modern estuarine crocodiles.

Figure 7.10. Skull and mandible of a Goniopholis *sp. from the Dinosaur Park Formation in Dinosaur Provincial Park, Alberta. Note the deep notch at the junction of premaxilla and maxilla, which is a basal character of Neosuchia. Skull length approximately 50.0 cm. Specimen in the Royal Tyrrell Museum of Palaeontology, Drumheller, Alberta.*

Because they were so generally similar to modern crocodiles, why do we not consider the goniopholids as the logical ancestors of Eusuchia? Because despite their advanced appearance, goniopholids retained a full complement of key basal mesoeucrocodylian characters, including flat to amphicoelous vertebrae, an incomplete secondary palate, internal nostrils that were not enclosed by the pterygoid bones, and even vestiges of an antorbital fenestra. Given that they showed many derived characteristics while at the same time retaining these basal features, we must assume they were a separate sister group of the Eusuchia, derived from the same ancestry in the Late Jurassic, but off in their own evolutionary direction.

Presently, we do not know which mesoeucrocodylian clade is the direct ancestry of eusuchians, and various groups have been proposed by various authors. For example, Buffetaut (1982) and many others have suggested that the Atoposauridae, a group of small, short-skulled, long-legged neosuchians of the Late Jurassic, were the closest eusuchian ancestors. This assumption stems from their age, the vertebral make-up of some atoposaurids (which showed limited development of procoelous articulation), and their palates, which have the choane located far to the back, close to the eusuchian position. However, cladistic analysis by Benton and Clark (1988), which was based on many characters, disputed the close relationship between the atoposaurids and eusuchians. Currently, they are generally regarded as another possible sister clade, very close to the eusuchians. Norell (1989) and Clark (1996) considered the Lower Cretaceous genus *Bernissartia* (see discussion to follow) as the closest sister group to Eusuchia, whereas Steel (1973) classified the genus to be actually within the Eusuchia, and Brochu (1999) considered its relationships uncertain. An even earlier genus, *Hylaeochampsa,* also lies very close to the root of the Eusuchia and may be arguably classified as a close ancestor or the basal member of the clade, as will be examined below.

Eusuchia

After reading about the complex interrelationships and diversity of the basal neosuchian groups, it will be refreshing to learn that in current cladistic taxonomy, the eusuchia are generally accepted as a monophyletic group. There are uncertainties about the basal members of the clade, but there is no serious doubt that most Late Cretaceous to present day eusuchians can be attributed to three higher groups: the Alligatoroidea, Crocodyloidea, and Gavialoidea, all derived from common ancestry. The unifying characteristics of all eusuchians have been mentioned at various places in previous discussions. Collectively, they are as follows: choane located at the rear of the palate bounded entirely within the pterygoids, a full secondary bony palate, vertebrae that are markedly procoelous (i.e., a ball and socket with the socket pointing forward), and multiple rows of dorsal osteoderms, rather than two rows typical of the basal mesoeucrocodylians.

Two genera of neosuchians have some eusuchian characters that appear at a partially developed stage, which suggest they may be close to or at the base of the Eusuchia. Alternatively, they may be parallel groups of neosuchians that evolved these advanced features independently of the true basal eusuchians (which we have not yet found as fossils). These genera are both from Europe: *Bernissartia* and *Hylaeochampsa*. *Bernissartia* has already been discussed as a possible ancestor of the eusuchia because it had imperfectly developed procoelous vertebrae and incompletely pterygoid-bound choane. It was a small (~1.0 m), blunt-toothed neosuchian, of typical aquatic crocodyliform proportions, known from only two skeletons (discovered and described in the 19th century) in the Early Cretaceous of Belgium. The skull of *Bernissartia* is typical of modern, broad-snouted crocodylians (i.e., like alligators and crocodiles), and the rostrum has the characteristic notch at the junction of the maxilla and premaxilla to accommodate the fourth dentary tooth. Overall, *Bernissartia* seems to be a reasonable candidate for the basal eusuchian, with some incompletely evolved characteristics. But unfortunately, there are not enough specimens and body parts known to determine with certainty whether it lies at, near, or parallel to the ancestry of Eusuchia.

Hylaeochampsa is an even earlier neosuchian, known from a single, toothless Late Jurassic skull from England. The specimen extrapolates to a small animal of 2.0 m overall length, with large alveoli (tooth sockets) that suggest it had blunt, crushing teeth, like alligators and *Deinosuchus*. A phylogenetic analysis by Wu et al. (1996) suggested that *Hylaeochampsa* is not only a eusuchian, but it also may be classified within the higher branch of alligatoroids. If this argument is correct, than the common ancestor of alligatoroids and crocodyloids would have to date back before *Hylaeochampsa*—that is, before Late Jurassic age. Brochu (1999) urged caution in this interpretation because it would engender over 50 million years of "ghost lineages" (i.e., unknown clades) to be accounted for between *Hylaeochampsa* and the first indisputable eusuchian clades known in the fossil record. An alternative view on the nature of *Hylaeochampsa* suggests it is an independent neosuchian lineage with some convergent features with eusuchians, especially in the palate and nostrils. This interpretation is a closer fit with the known fossil record of Eusuchia but requires better material to confirm or deny it.

The earliest unmistakable eusuchians are of Late Cretaceous age and include *Deinosuchus, Leidyosuchus* (discussed in Chapter 6), and several crocodylians with obviously alligatoroid characteristics. In fact, so many eusuchians appear in the Late Cretaceous within a brief geological time that it seems a burst of evolution occurred once the eusuchian adaptations had appeared. This may result from the improved feeding and breathing advantages afforded by the complete secondary palate and posterior nostrils of Eusuchia. Or, as argued by Busbey (1997), the extraordinary strength of the rostrum resulting from the internal buttressing by the secondary palate may have given Eusuchia of the Late Cretaceous the edge they needed to compete in a world of huge dinosaurs.

At this point in the discussion, we may resurrect the term "croco-dylian" and use it in the strictly correct sense. Benton and Clark (1988) first defined the modern concept of the Crocodylia as those eusuchians with scapulae (shoulder blades) that do not broaden greatly in the dor-sal direction. This seemingly trivial distinction separates the three high-er modern groups (alligatoroids, crocodyloids, and gavialoids) from *Bernissartia* and *Leidyosuchus* of the Cretaceous. However, a later ana-lysis (Brochu 1997a) showed that some *Leidyosuchus* species belong within the Alligatoroidea and thus are crocodylians in the strict sense. Because *Deinosuchus* is an alligatoroid, I will continue the pattern introduced in this chapter and discuss first the characteristics of the other crocodylian clades, then end the chapter with discussion of the phyletic position of *Deinosuchus* and the other alligatoroids.

One of the enduring controversies in taxonomy of the Crocodylia has been the relationship of the modern and fossil "gharials" of the genus *Gavialis* (using the common Indian name for the younger forms) with the living "false gavial" *Tomistoma* and with the other croco-dylians. These genera are the modern eusuchian longirostrines (i.e., long snouts), with skull proportions similar to some of the more primi-tive groups we have discussed, such as the teleosaurs. Their extremely narrow, elongate rostrums tend to make them appear very different from the broad-snouted (brevirostrine) crocodylians, and earlier as-sumptions were that the false gavials may trace their ancestry to Late Cretaceous longirostrine eusuchians, such as *Thoracosaurus* (e.g., Steel 1973), whereas the gharials originated in the Tertiary. However, as will be discussed, it is evident that the longirostrine morphology is a repeat-ing theme in crocodyliform evolution (as is gigantism), which reevolves in various groups as an adaptation to fish eating, and the modern false gavial may or may not be closely related to any Cretaceous genus.

The modern gharial, *Gavialis*, is a slender, freshwater genus com-mon in rivers in northern India, and it is the best-known example of modern longirostrines. They reach lengths reliably documented at 6.0 m and are abundant in the Ganges River, where *Gavialis gangeticus* may opportunistically feed on the remains of dead people deposited ritually in the sacred river. (Gharials are clearly adapted as fish eaters, and even the largest individuals are normally not predators or scaven-gers of tetrapods as large as humans.) Fossil species of *Gavialis* are known as far back as the Miocene Epoch, chiefly in India and adjacent Asia. The false gavial, *Tomistoma*, is a smaller crocodylian (~4.0 m) of generally similar proportions, found in freshwater habitats in Malay-sia. Its ancestry has been traced far back into the fossil record in older literature, but modern genetic comparisons suggests that these assump-tions are incorrect and that it actually shares the same ancestry as *Gavialis*. There are some notable skull differences between the true and false gavials: in *Tomistoma*, the orbits are large and nearly continuous with the temporal fenestrae. The snout broadens slightly at the rear of the rostrum, and the skull table (i.e., the flat, posterior dorsal surface) is quite small. In *Gavialis*, the rostrum is nearly parallel-sided behind the premaxilla, the orbits are smaller, and there is a substantial postorbital bone between the orbits and the supratemporal fenestrae. In addition, *Gavialis* has a distinctive pattern of osteoderms, with the number of dorsal rows changing from six to four as the individuals mature (Ross 1989). Although these differences have long been considered to be important signs of their distant relationship, comparison of DNA from living *Tomistoma, Gavialis,* and a range of other modern crocodylians (Hass et al. 1992), indicates that the two living longirostrine genera are quite close—and not all that different from the brevirostrines either.

But there is a notorious "gharial problem" concerning the phylogeny of the narrow-snouted living crocodylians (Brochu 2000). Even though *Gavialis* and *Tomistoma* are quite close genetically, phylogenetic analysis that is based on many characters produces cladograms that show the living false gavial is actually close to the true crocodiles (i.e., the Crocodyloidea), whereas the true gavials should be placed into a separate higher group, currently termed the Gavialoidea. It is not evident which classification approach is valid because they cannot both be correct, but fortunately for our purposes, neither longirostrine group is in the linear ancestry of *Deinosuchus,* and we may therefore move on to the Crocodyloidea.

As with so many other discussions here, there are some anomalies relating to the crocodyloids. This clade as defined by Norell (1989) includes modern and fossil members of the typical crocodile genus, *Crocodylus,* and its sister group, the modern dwarf crocodile, *Osteolaemus.* Some cladistic analyses of higher crocodylian relationships also place the false gavial, *Tomistoma,* in the Crocodyloidea, but as discussed previously, the placement of *Tomistoma* is highly controversial. The most easily recognized characteristic of the crocodyloid clade is their moderately broad snout, which tapers anteriorly to a prominent notch at the contact of the maxilla and premaxilla (into which the fourth dentary tooth inserts). In addition, true crocodiles have relatively pointed teeth, with many of the anterior series exposed when the mouth is closed, and they possess salt-secreting glands in the oral cavity (Taplin and Grigg 1981), as well as salt-excreting abilities in the cloaca (the combined excretory organ). Among living crocodiles, the range of variation in form is modest; at one end of the range is the smaller, slender-snouted species *Crocodylus cataphractus* of Africa, which seems to be both the most primitive of living species (Brochu 2000) and also the closest within the genus to the dwarf crocodile *Osteolaemus.* Among other characters, *C. cataphractus* has a distinctive pattern of nuchal (neck) osteoderms and subtleties of the rostrum reminiscent of *Tomistoma.* Fossils of *C. cataphractus* are found in the Pliocene of Africa, showing that the species is at least a few million years old. At the other end of the range are the larger, broad-snouted, shorter-skulled species, such as the Cuban crocodile, *C. rhombifer,* and the African Nile crocodile, *C. niloticus,* which is also the type species of *Crocodylus.* Steel (1973) listed 66 species of *Crocodylus,* including living and fossil forms. And although it is clear that many of Steel's species are highly doubtful (some species are based on only a few bones or teeth), there are still a lot of obviously valid species among that large number. Many of the described species of *Crocodylus* are alive today, and some range back in the fossil record to the Miocene Epoch. In the modern world, we observe some *Crocodylus* species overlapping in the same approximate geographic range but in different habitats within those areas, which may explain why so many species of a single genus could exist in the past and present. For example, in the general Caribbean region, from Florida to the northern coast of South America, one may find *Crocodylus acutus* (the American crocodile), *C. rhombifer* (the Cuban crocodile), and *C. moreletii* (Morelet's crocodile) in various subhabitats. Similarly, six species of *Crocodylus* (*C. porosus, C. johnsoni, C. novaeguineau, C. raninus, C. siamensis,* and *C. mindorensis*) overlap in the general area we call the South Pacific, from the Southeast Asian mainland to northern Australia. The many coexisting living species of the single genus *Crocodylus* contrast with the ranges of other large modern crocodylians, where a single species tends to dominate a large area. For example, *Alligator mississippiensis* is the sole crocodylian in North America (aside from the southern tip of Florida) and *Gavialis gangeticus* dwells alone in parts of northern India. *Crocodylus* seems to evolve multiple species in tropical subhabitats quite easily, perhaps because they are highly adaptable, opportunistic feeders.

Because modern *Crocodylus* species fit the most closely into the popular concept of what a "standard" crocodyliform should be, it has commonly been assumed that the group must be an old, basal stock of eusuchians traceable to the Late Cretaceous (e.g., Taplin and Grigg 1989). However, modern cladistic analysis of the fossil record shows otherwise; the oldest true crocodiles seem to be traceable only back to the Miocene Epoch. An Oligocene species, *Euthecodon arambourgi* from Africa, was indicated by Brochu (2000) to be the closest sister taxon to the crocodiles. If we accept this assumption, then older fossils identified as species of *Crocodylus* from the Eocene of Europe and western United States (Markwick 1998b; Steel 1973) cannot have been properly placed within the genus. Further, some much older fossils attributed to *Crocodylus,* especially those of the Late Cretaceous, were obviously incorrectly assigned to genus.

These misidentifications largely result because many fossil crocodylians possess the same general skull shape (with the premaxillary–maxillary tooth notch) as true crocodiles, as opposed to modern alligators that lack this notch. But the tooth notch turns out to be a primitive feature of all the Neosuchia, rather than a derived feature of the true crocodiles. Some Cretaceous eusuchians, including *Leidyosuchus* and *Deinosuchus,* have the overall skull shape of *Crocodylus* because their skulls possessed the same notch; but ironically, these genera turn out to be closer to the ancestry of modern alligators rather than to that of crocodiles. The notably broad snouts and absence of tooth notches in modern alligators are derived features of the higher branches of the alligatorid family, and these distinctive features tend to camouflage the antiquity of the larger alligatoroid group.

In fact, alligator ancestry seems to be very old and may lie at the base of the Eusuchia. We have fossil alligators known from the Late Cretaceous, and there is little dispute that these are indeed early branches off the lineage leading to modern alligators and caimans. Further, and notably, current cladistic analyses suggest that virtually all Middle and Late Cretaceous eusuchians known to date are alligatoroids. As designated by Norell et al. (1994), who resurrected a 19th-century name, the Alligatoroidea is the higher grouping of crocodylians defined to include the derived alligator family (Alligatoridae; in the vernacular, alligatorids) and their stem group ancestors. Because the modern alligators are a fairly uniform set of very broad-snouted crocodylians, consisting of only two species of *Alligator* and five species of caimans, one might expect fossil alligator relatives to be similarly limited in diversity. But the alligatoroids present us with exactly the reverse situation of the crocodyloids, where the modern crown group of crocodiles had only a limited set of fossil taxa in their ancestry. As we trace back the ancestry of the alligators, we face an evolutionary bush of diverse forms, many of which barely resemble the derived alligators except in several subtle anatomical features.

The characteristics that unify alligatoroids in current cladistic analyses are all details of skull morphology (Brochu 1999; Norell et al. 1994). Two of the more easily recognizable are sutural patterns. In alligatoroids, the suture that bounds the maxillary and ectopterygoid bones diverges medially away from the posterior teeth, whereas in crocodyloids, it runs parallel to the tooth row. Also in alligatoroids, the suture between the surangular and angular of the lower jaw forms a long ventral extension, whereas it is fairly simple in crocodyloids. These, and other details of the bone contacts and shapes, may seem to be trivial bases for determining higher taxonomic relationships. But when they consistently co-occur in obviously related animals, it suggests they are valid diagnostics when applied to less obviously related forms. For example, I have observed that the two sutural features noted above are present in the alligatoroid condition in *Deinosuchus*, which tends to reinforce its higher affinities on the basis of dental characteristics. Because it is highly improbable that these characteristics all arose in the same direction independently, I accept that they are sufficient reason to place *Deinosuchus* in the Alligatoroidea.

Among the less definitive diagnostics of alligatoroids, we observe they are commonly blunt-toothed, sometimes extraordinarily so. In some alligatoroids, such as the Cretaceous genus "*Bottosaurus*" (which may be an invalid name for an eastern United States form of *Brachychampsa*) and the Early Tertiary *Allognathosuchus*, the posterior teeth are of the so-called globidentine shape, which ranges from almost flat to hemispherical. (In contrast, the gavialoids and crocodyloids tend to have mostly pointed teeth.) There are exceptions in both directions, especially with some alligatoroid teeth being quite pointed; but many more blunt-toothed forms occur among the alligatoroids than the other Eusuchia. Further, there is a tendency for alligatoroids to have most (or all) of their teeth contained in occlusion pits within the jaws when they are closed, rather than having the tips exposed. This characteristic of tooth occlusion in pits is quite variable among eusuchian species and occurs to some degree even among the crocodyloids, but it is present to a significant degree in most alligatoroids.

Other distinctions among higher crocodylian groups are not always evident in fossils. Among living eusuchians, the alligatorid family lacks the salt-secreting glands in the mouth and cloaca that are present in living crocodiles, as discussed previously. It is parsimonious to assume that the crocodyloid-type adaptations evolved only once, and if this assumption is correct, then all alligatoroids had to overcome the effects of saltwater immersion by other means, such as by having low-permeable skin or highly efficient kidney function. Neither of these properties would likely be preserved as fossil evidence, so they are simply speculations; nevertheless, Brochu (1999) pointed out that the caimans of South America and the Chinese alligator somehow made sea crossings to reach their present locations. Therefore, they had to possess some means to handle the saltwater problem, as was discussed in Chapter 6 concerning the hypothetical crossing of the Western Interior Seaway by *Deinosuchus*.

As the longest-lived clade of eusuchians, the Alligatoroidea contains a broad range of forms, including the largest known crocodylians and some odd skull shapes. However, the oldest alligatoroids are not strikingly odd. Some resemble modern alligators, and some appear to be more typical basal crocodylians. In the Late Cretaceous of western North America, there are several smaller alligatoroids (<2.0 m) that are known from excellent specimens. These include *Albertochampsa*, from the Campanian of Alberta (Fig. 7.11), and *Brachychampsa*, from the Maastrichtian of Montana. Both had broad, rounded snouts that resemble derived alligators; however, Norell et al. (1994) analyzed a large suite of morphological characters of *Brachychampsa montanensis* and found that they were not members of the crown group alligatorid family despite their superficial resemblances. These early alligatoroids lived in freshwater habitats in the western United States, and closely related or not, they probably filled essentially the same ecological role as smaller modern alligators. Also in the Late Cretaceous is *Leidyosuchus canadensis*, the type species of the genus (see also p. 121), which Brochu (1997a) classified as a basal alligatoroid. As discussed previously, *Leidyosuchus* was a broad-snouted eusuchian with a tapering skull, the basal neosuchian notch at the junction of the premaxilla and maxilla, and mostly pointed teeth. But despite its lack of externally alligatoroid characters, Brochu found a sufficient number of skull and postcranial characters to classify the type species of *Leidyosuchus* as an alligatoroid. *Leidyosuchus* was apparently very successful because fossils attributed to the genus ranged through the Cretaceous–Tertiary extinction boundary and up into the Eocene (at which age they were very abundant in the western United States). Brochu (1997a) also replaced several species originally assigned to *Leidyosuchus* into a new genus, *Borealosuchus*, which he tentatively classified as a sister group of all three higher crocodylian groups—that is, not readily classified as crocodyloid, alligatoroid, or gavialoid, but rather close to the base of all three.

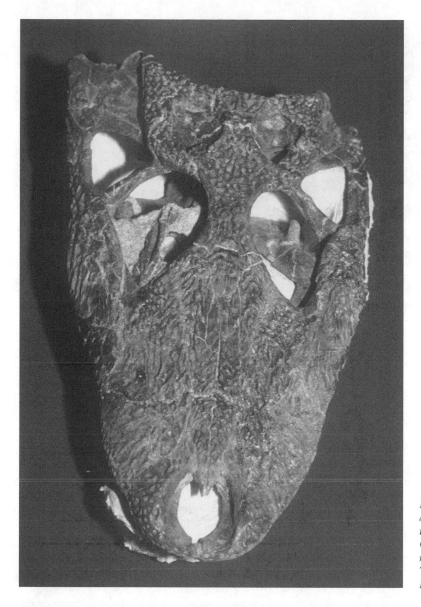

Figure 7.11. Skull of an early alligator, Albertochampsa, *from the Upper Campanian of Alberta, Canada. Skull length approximately 20.0 cm. Specimen in the Tyrrell Museum of Natural History, Drumheller, Alberta.*

Deinosuchus and the Alligatoroidea

Deinosuchus is quite evidently a genus that we can classify among the basal Late Cretaceous alligatoroids. Some of the features discussed in the previous section can be easily demonstrated; for example, the posterior teeth of *Deinosuchus,* including the heavy makeup of the rounded rear teeth, are consistent with the characteristics of alligatoroids. Regarding tooth occlusion, *Deinosuchus* had nearly all lower tooth tips (except the enlarged fourth mandibular tooth) occlude into

Figure 7.12. *Occlusion pits in the maxilla of* Alligator mississippiensis. *Compare with Figure 6.4, which shows similar pits in the maxilla of* Deinosuchus.

pits located just inside the tooth row of the maxilla and premaxilla (see Figs. 6.4 and 7.12). It is evident that when a *Deinosuchus*' jaws were locked together, all the lower teeth were hidden, except the fourth on the dentary. The upper jaw teeth were probably partly exposed because they passed labially (i.e., outside) over the dentary teeth and fit snugly against the outer surface of the lower jaw. It appears that most upper teeth had their tips slightly exposed, unless there was a thick skin fold on the lower jaw covering them (which is undeterminable from the fossils).

Brochu (1999) observed that despite much incompleteness of our knowledge of *Deinosuchus,* we are aware of several additional features that clearly place it in the alligatoroids but near the very base of the group. One subtle alligatoroid detail of *Deinosuchus* I can identify from the single ilium attributable to the genus (Fig. 7.13). It was collected in western Georgia, less than 45 km from where I am writing this book, in association with a *Deinosuchus* mandible. Assuming this ilium comes from the same animal (and the size is correct for the association), it confirms that the alligatoroid character of having a high, smooth, posterodorsal margin on the ilium was present in *Deinosuchus*. This feature contrasts with the condition in crocodyloids, where the posterior iliac margin is depressed or strongly indented. (The iliac margin is also high in gavialoids, but there is little chance that *Deinosuchus* is related to the gavials, so the separation from crocodyloids is far more significant.)

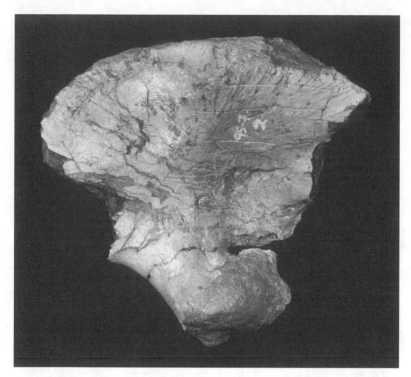

Figure 7.13. Ilium associated with the Deinosuchus *mandible from the Blufftown Formation in Chattahoochee County, Georgia (see Fig. 4.7). Although it is fragmentary, the posterior margin (toward left) is high and smooth, as in alligatoroids, rather than indented, as in crocodiles. Width of specimen as shown is 10.0 cm, corresponding to a smaller* Deinosuchus *~6.0 m in length.*

A revealing basal alligatoroid characteristic of *Deinosuchus* is the tendency to have the third and fourth teeth in the dentary located in joined or confluent alveoli (i.e., where two teeth emerge from a virtually single socket). This feature is not absolutely present in all *Deinosuchus* specimens in which the anterior dentary region is preserved, but it has definitely been observed in specimens from both Texas and North Carolina, and therefore it is present in both western and eastern populations. The same conjoined dentary tooth character is observed in some other primitive alligatoroids and eusuchians close to the alligatoroid base, notably *Leidyosuchus* and another basal eusuchian genus, *Diplocynodon* (which means "double canine tooth"). This latter genus is known from many species in the Early Tertiary of Europe, and in addition to the odd doubled teeth, it is diagnosed by having ventral osteoderms that are composed of two bony units in association (rather than the single ossifications typical of most reptiles). *Diplocynodon* seems to be a sister taxon of the alligatoroids (that is, a separate group but close to the ancestry of the alligatoroids), and it may closely share the same ancestry as *Deinosuchus*.

Indeed, on the basis of the collection of both basal and derived characters they possessed, it is possible that *Deinosuchus, Leidyosuchus,* and *Diplocynodon* formed a natural group that branched off from the mainstem of alligatoroid evolution early in its history, and certainly by Campanian Age, when both *Leidyosuchus* and *Deinosuchus* were present. It is also likely that other Late Cretaceous alligatoroids, such as *Brachychampsa*, were closer to the branch of the Alligatoroidea that ultimately led to the true alligators. If these are correct assumptions, then the stem leading to modern alligatorids was established by the Campanian Age too because the alligatorlike genera *Albertochampsa* (Fig. 7.11) and *Brachychampsa* (Williamson 1996) were present at that time. A final conclusion, therefore, is that *Deinosuchus*, although a basal alligatoroid, has no close or linear relationship with modern *Alligator*.

Before leaving the topic of the alligatoroids, we should note some of the extraordinary animals that have emerged from the clade during its long history. Brochu (1999) compiled a sizable cladistic analysis of the Alligatoroidea and analyzed some of the morphological characters that tended to recur through their history. Among the oddities of alligatoroid evolution are globidentine-toothed, broad-snouted (sometimes virtually duck-billed), and short-snouted forms, as well as several gigantic species. Some of these characters run together: for example, *Deinosuchus* had slightly globidentine teeth and a premaxilla sufficiently broad to come close to being duck-billed, and it was clearly gigantic. But some of the oddest alligatoroids were derived caimans from the Late Tertiary of South America.

In Chapter 3, I discussed *Purussaurus*, which is among the largest known crocodylomorphs of all time, equaling the skull length of *Deinosuchus* and probably exceeding slightly its bulk. *Purussaurus* was a gigantic caiman and thus a member of the alligatorid family, with a strangely depressed nasal region and a broad snout. Remarkably, there is at least one other caiman species of equal size, with skulls that are even longer than those of *Purussaurus* and *Deinosuchus*, although proportioned differently. These also come from the Late Tertiary of South America and may be closely related to *Purussaurus*. This odd giant caiman is named *Mourasuchus atopus* (Langston 1966) and features a skull with a long, parallel-sided rostrum terminating in a blunt snout. In addition, it has oddly slender, lightly built lower jaws, with many small teeth, and the posterior bones of the skull are greatly reduced. From the skull length, we can extrapolate that *Mourasuchus atopus* had a body length of ~12.0 m, but it is difficult to imagine what such a huge crocodylian was feeding on with such a weak jaw and small teeth. To presage the next chapter, it is fortunate that with *Deinosuchus*, we can make excellent inferences about its feeding behavior as related to its great size, with some of our inferences based on direct evidence.

8. The Prey of Giants

Signs of Ancient Predation

A gigantic crocodylian, living in an age of huge dinosaurs, naturally leads to assumptions and questions about who was eating whom. Even the first description of *Deinosuchus* (*Phobosuchus*) *riograndensis* (Colbert and Bird 1954) contained explicit speculations that *Deinosuchus* evolved its great size the better to prey upon dinosaurs. In fact, Colbert and Bird listed a variety of herbivorous dinosaur species as possible prey of *Deinosuchus,* including "trachodonts" (an older term for flat-headed hadrosaurs), ceratopsians, sauropods, and armored dinosaurs (i.e., ankylosaurs). However, not all of their proposed prey species could have been *Deinosuchus* food. Better stratigraphic data from subsequent years of study in Big Bend shows that the only Late Cretaceous sauropod of the region (and in fact in all of the western United States) was *Alamosaurus sanjuanensis*. But *Alamosaurus* occurs in the Javelina Formation, which is the next rock formation up above the Aguja Formation, and does not contain *Deinosuchus*; therefore, *Deinosuchus* and sauropods apparently did not coexist in time. However, all the other dinosaurs mentioned by Bird and Colbert were indeed in the Aguja Formation and potentially associated with *Deinosuchus*— and thus the general contention of a likely predator–prey association remains valid.

This chapter contains three sections: the first on the general nature of fossil evidence of predation; the second on my observations that the southeastern *Deinosuchus* population fed heavily on marine turtles; and the third on the idea of dinosaur feeding by *Deinosuchus*. The last two sections divide roughly by geographic regions because the huge western deinosuchids were the most likely dinosaur predators, and all the evidence of turtle-eating is known from the east. However, the dichotomy is not all that precise because we have both direct and

Figure 8.1. In pursuit of a marine turtle. Drawing by Ron Hirzel.

inferential evidence that some dinosaur predation by *Deinosuchus* (of an unexpected type) was occurring in the eastern continent. And there is no particular reason to assume that the Big Bend giant crocodylians fed exclusively on dinosaurs and avoided the turtles found as fossils in the Aguja Formation.

When paleobiologists attempt to reconstruct the feeding behaviors of extinct animals, there are several directions to search for information and evidence. Clearly, we start by assuming that extinct animals behaved roughly like their living relatives. With that in mind, the initial assumption is that *Deinosuchus,* as a huge crocodylian, was a top predator capable of feeding on large prey, with some victims approaching its own size. In addition, from observations of modern brevirostrine crocodylians, especially *Alligator* and *Crocodylus* species, it is clear that they prey on an unusually wide variety of animals and that their favorite foods change as they mature and increase in size. They may also seek new prey with the change of seasons and as opportunities arise, such as when a migrating herd of wildebeests in Africa present themselves at a water hole for a *Crocodylus niloticus,* or when an abundance of snapping turtles or a hatching of waterbirds are available for an *Alligator mississippiensis.*

The most direct evidence of prey selection in any extinct species is to find fossil remains of the stomach contents or gut contents (the latter termed "cololites"). Cololites are apparently fairly common (Seilacher et al. 2001), whereas actual fossilized stomach contents are rare. But when it is available, such evidence is about as clear and unambiguous about ancient feeding behavior as any data can be in the fossil record. We have available such remains in a few Cretaceous vertebrates; for example, the gut contents were found in a fossil shark, *Squalicorax falcatus,* from the Late Cretaceous of Kansas (Druckenmiller et al. 1993). The remains included pieces of some larger marine animals (mosasaurs), suggesting the shark had been scavenging rather than preying upon them (Schwimmer et al. 1997b). Likewise, an armored dinosaur, an herbivore, from the mid-Cretaceous of Australia, was reported by Molnar and Clifford (2000) to contain a small amount of plant material in the approximate gut region. To date, the fossil record of Mesozoic land-carnivore stomach contents (and cololites) is nil, and the general quality of *Deinosuchus* fossils is far too poor to expect that such a wonderful specimen might appear.

Another fairly direct line of evidence of predation continues down the digestive system with the study of the preserved feces (termed "coprolites") of an animal. Here, we would search for undigested, identifiable bits of the prey in the background mass of the coprolite. Coprolites are surprisingly common from certain vertebrate groups, especially sharks and bony fish, where they tend to be of distinctive appearance (Fig. 8.2) and are well preserved in phosphatic marine sediments. Coprolites of Cenozoic mammals are also quite common, especially in such areas as the Oligocene deposits of the Big Badlands in South Dakota, where the abundant oreodonts (small, herbivorous hoofed mammals) left masses of coprolites. (However, Seilacher et al. 2001 have argued recently that these very fecal-looking masses are actually cololites!) In the Cretaceous, the coprolites of herbivorous dinosaurs have been extensively studied (e.g., Chin 1997; Chin and Gill 1996), with a great amount of information available about the food of whichever species produced the feces. But that "whichever" reflects the difficulty in determining the precise source of a coprolitic mass. Chin (1997) noted that very large animals' coprolites are preserved with difficulty because they would be "mechanically disrupted" (i.e., trampled), and Chin and Gill (1996) showed that they attracted dung beetles even as far back as the Mesozoic. Chin also discussed the difficulty of assigning precise origins to the coprolites of Cretaceous carnivores because they might come from a range of predators, including theropod dinosaurs, large turtles, and crocodylians. Nevertheless, she and others (1998) identified a huge coprolite from the Maastrichtian of Montana as that of a *Tyrannosaurus rex.* Within that remarkable specimen, there was a large percentage of bone chunks, including many identifiable as coming from an ornithischian dinosaur.

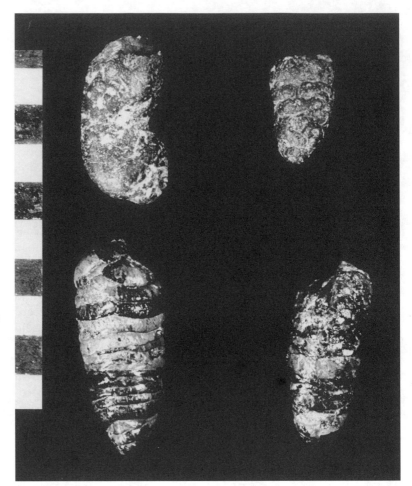

Figure 8.2. (above) Coprolites, probably from bony fish, from the Blufftown Formation at Hannahatchee Creek, Georgia. All of the coprolitic material from this formation is replaced by apatite.
(below) Shark coprolites from the Mooreville Formation, found in association with the Deinosuchus *specimen from the Alabama River site, Lowndes County, Alabama. The distinctive spiral shape is diagnostic of the intestinal spiral valve of selachians.*
Photographs by Tracy Hall.

The question naturally arises, "Do we have coprolites of *Deinosuchus*?" The answer is, "Maybe, but they won't help answer the larger question of what they ate." Among the *Deinosuchus* fossils and sedimentary materials in the Blufftown Formation beds of Georgia and Alabama, I have been collecting possible coprolites for many years. Because the Blufftown Formation is a coastal marine deposit with many *Deinosuchus* teeth and other remains (Chapter 5, p. 89), I would expect to find preserved fish and shark feces in it (Fig. 8.2), which are often preserved in the phosphatic sediment. But I have also collected several coprolitic-looking masses that are too large to be attributed to any known fish or shark in the formation; they may have come from either a big crocodylian or another large marine tetrapod (e.g., a mosasaur, plesiosaur, or a very large sea turtle). These masses are of two general forms. The first type are elongate, tapering at each end, and very irregular, with bits of small bone fragments throughout (Fig. 8.3). The other coprolites are tubular, divisible into irregular disklike units, and with a uniform internal texture of fine material showing no apparent

Figure 8.3. Possible Deinosuchus *coprolites from the Blufftown Formation at Hannahatchee Creek, Georgia. (right) Alternative type of coprolite in* Deinosuchus *deposits, showing a definite overall form of sausagelike units terminated by possible sphincter imprints, but without discernible internal bone material.*
(left) Larger coprolitic form of amorphous (but generally consistent) overall shape, containing numerous bits of bone. Scale in centimeters. Photographs by Tracy Hall.

bone fragments. Modern crocodylians emit feces with poorly defined shapes and contents that vary, depending on their food. They often have bits of fish scales in their feces if they are feeding on gar and other rough-scaled fish, and they also feature bits of undigested bone. In considering the end products of digestion in modern crocodylians, one must remember that they actively dismember their prey, but their pointed or rounded teeth do not easily allow meat and bone to be comminuted into fine bits. (This contrasts with the prey processing of carnivorous mammals, which have specialized molars such as carnassials that slice muscle tissue efficiently, and bone-crushing molars such as those of hyaenids and canids.) On the basis of the comparison with modern crocodylians, the irregular coprolites in the Blufftown Formation would seem to be most likely attributable to the digestive processes of a crocodylian. Unfortunately, to date, none of the bits of bone I have found in such masses are identifiable, except to be sure that they are not fish scales. Therefore, even if I have identified *Deinosuchus* coprolites (which is by no means certain), they have not yet revealed prey information. So far, no coprolites attributable to *Deinosuchus* have been reported from any other localities in the east or west.

However, some indirect digestive evidence may be at hand to bear on the topic. In a larger discussion of Late Cretaceous events on the Atlantic Coastal Plain (Gallagher 1993), there was a brief mention of carnivorous dinosaur teeth from a marine deposit in New Jersey that apparently had their enamel surfaces partially dissolved away (discussed in detail below). Gallagher's assumption was that the enamel of these teeth had been stripped away in the digestive juices of a crocodylian. To carry this assumption further, *Deinosuchus* is also present in the same locality (the Ellisdale site in the Marshalltown Formation; see Chapter 5, p. 82), and because of its size and tooth morphology, we must assume that if a crocodylian was feeding on a theropod dinosaur, it was a *Deinosuchus*. It may seem a bit surprising to consider that among all the possible dinosaur prey, *Deinosuchus* would feed on a carnivore; but several lines of evidence support the assumption. First, we may observe modern *Crocodylus niloticus* in Africa occasionally preying on carnivores as large as lions (Cott 1961). If we extrapolate the relative sizes of a modern large *Crocodylus* to a lion, it becomes clear that even a smaller eastern *Deinosuchus* (of, say, 8.0 m length) would be capable of killing and dismembering a theropod of well over 1000 kg in weight. There is indeed no reason that one carnivore may not opportunistically prey on another, especially on smaller, younger animals. A second line of evidence supports Gallagher's contention that crocodylians were feeding on theropods: there is a theropod bone bearing evidence of having been chewed and crushed by a large crocodylian. That specimen and related matters of tooth marks will be discussed subsequently.

Another, still less direct source of information on the predatory behavior of a carnivore comes from the morphology of the animal in study, and especially its teeth. In the case of *Deinosuchus,* we know a great deal about the teeth, and they are indeed suggestive of specific feeding preferences. The teeth of crocodylians may vary greatly; at one extreme are the long, thin, pointed teeth of gharials and other long-snouted forms, which are adapted to catching fish. At another extreme are the complexly denticulate teeth of several Mesozoic short-snouted mesoeucrocodylians (see Chapter 7) known from Africa and Madagascar, apparently adapted for feeding on vegetation. Among living broad-snouted crocodylians, we observe a fairly conservative range of tooth shapes, with crocodiles possessing mostly stout, conical, pointed teeth and a few blunt teeth in the rear of the jaws, and alligators and caimans having a mixture of the same stout, conical teeth and many rear blunt teeth. A related observation is the occlusion (i.e., interlock on closure) pattern of typical *Crocodylus* versus *Alligator* species. In the former, the anterior teeth interlock with enough precision to cause a jagged tear that will sever a body part that has been cleanly bitten. In contrast, *Alligator* has less shearing effects from closure of its anterior teeth, but also has more effective crushing by teeth in the rear jaws. We can observe the relationship of these dental patterns to their choices of prey. Crocodiles, when mature, attack medium-sized and even large mammals and reptiles, in which cases the severing of large body parts becomes necessary in order to consume the animals. In contrast, larger alligators tend to feed on a range of fish, amphibians, reptiles, birds, and small and medium (small deer size) mammals. It is commonly observed here in the southeastern United States that medium and large alligators tend to feed heavily on freshwater turtles and that the blunt rear 'gator teeth do a good job of crushing the shells. I would guess that the more pointed teeth of *Crocodylus* species would be more likely to pierce (but not crush) the shells of larger turtles and would not be as effective for this form of predation.

The teeth of *Deinosuchus* are not all that grossly different in form compared with those of alligators, except in being proportionately shorter and more rounded in the middle and rear portion of the jaws and in being unusually thick in all positions. As we have considered in Chapter 2, the extremely heavy, slightly infolded enamel of the teeth and the thick interior dentine layers were first described by Emmons (1858), and it was this makeup that led him to coin the species name "*rugosus*" (meaning "rugged" or "rough") for the bearer of such teeth. Carrying this idea further, and assuming that nature is parsimonious in allowing animal structures to evolve and persist, we must conclude there was an adaptive function for these heavy teeth in *Deinosuchus*. That function can only have been for crushing the bones and hard parts of larger animals with which deinosuchids coexisted. As I will discuss at some length to follow, it appears that the prey choice of the smaller eastern deinosuchids commonly took best advantage of the thick teeth by employing them to crush the heavy shells of the common nearshore marine turtles (Schwimmer and Williams 1996). If we carry that idea further, it leads to speculation that the heavy tooth morphology of *Deinosuchus* in the smaller eastern animals preceded and selected for the huge size reached by deinosuchids in the west, when they applied those bone-crushing teeth primarily to big ornithischian dinosaurs. This idea will be pursued, along with the evidence, in the last section of the chapter.

Continuing down the sequence of predatory evidence, we come finally to evidence of feeding left behind on the prey itself. In the modern world, we can obviously watch the process of animal feeding directly, and frequently we can find prey that have survived a predatory attack and lived with the signs preserved. For example, in a small lake in nearby eastern Alabama live both a large snapping turtle and a smaller (2.0 m) alligator, each missing part of a hindlimb. Although no one saw the cause of those events, the same lake hosts a ~3.0 m alligator, whose mass exceeds twice that of the smaller alligator, and it is fairly obvious that the larger 'gator caused the loss of limbs in the other animals. By the same reasoning, we may consider the makeup of a fauna and signs on some fossil prey, and then reconstruct ancient predatory activity. The best evidence to use in this work is bite marks and deformities on bones, and then to consider which species in that assemblage has the teeth and behaviors most likely to be the cause of those marks. In the next section, I will describe and illustrate some bite marks that may be attributed to *Deinosuchus* appearing on bones from several types of animals in the eastern Campanian localities, and on dinosaurs in the west. And, by no coincidence, many of the *Deinosuchus* teeth found in the east show the results of chewing on hard prey, which point to one specific set of predator–prey interactions.

A Feast of Eastern Sea Turtles

When I began working in the Late Cretaceous deposits of western Georgia and nearby Alabama, the first general observations I made were about the abundance of turtle bones and big "crocodile" teeth (as I perceived them at the time, not then realizing that *Deinosuchus* was an alligatoroid) in my routine fossil collections. Because the modern predator–prey relationship of alligators and freshwater turtles is well known, I immediately considered that there might be a similar relationship reflected in these fossil abundances. Initially, this was just speculation without direct evidence; but nevertheless, the coincidence was compelling.

In the Campanian-age fossil beds of the eastern Gulf Coastal Plain along the Georgia–Alabama border (see p. 88 for a discussion of localities), the remains of sea turtles are truly superabundant. By far the majority of these Cretaceous turtle bones come from a single species named *Bothremys barberi,* which was of the suborder Pleurodira (i.e., a "side-necked" turtle), in the large family of the Pelomedusidae. Side-necked turtles still exist and are fairly common, especially the pelomedusids, but they are restricted to the Southern Hemisphere and are of very low diversity, with only a dozen or so living genera. During the Late Cretaceous, however, they were globally distributed and very abundant in North America. Many of the *Bothremys* fossils we find in the eastern United States indicate that individual animals had shell lengths of ~1.2 m, which is comparable to larger modern loggerhead and green sea turtles. The most common *Bothremys* fossils are pieces of the carapace and plastron (the dorsal and ventral shells, respectively), and they are so common in some Georgia fossil localities that we collect many kilograms a day of shell fragments, each chunk possibly representing a different animal's remains. Typically, the shell bone is 1.0 to 2.0 cm thick, and may incorporate the thicker remains of the ribs (which make up most of the carapace) or limb attachment areas around the hip and shoulder girdles. As I will discuss below, several of these *Bothremys* remains have revealed predatory bite marks that seem to be large crocodylian in origin.

Many other turtle fossils are common in the Late Cretaceous marine deposits along the eastern Coastal Plain, representing a wide range of families. Most of these other turtles are of the suborder Cryptodira (i.e., "hidden neck" turtles), which is the common group that includes nearly all modern turtles, both marine and freshwater. In the southeastern United States, we have identified the fossils of many marine cryptodires, including giant turtles of the family Protostegidae that had carapace lengths ranging to 2.6 m and a wide variety of primitive types in the extinct families Chelosparginae, Toxochelyidae, and Prionochelyidae (Zangerl 1953). We also find early representatives of the modern sea turtles (the Cheloniidae) among the Late Cretaceous fossils, and additionally, we frequently collect the pebbled-shell fragments of an unidentified species of Trionychidae. These last were a group of cryptodires that in the modern world are freshwater dwellers and are called "soft-shelled" turtles. They were (and are) not really soft-shelled because the carapace contains the same costal (i.e., rib) bones that make up the shell of other turtles; but in the living forms, they are covered by a leathery skin rather than the thin, hard scutes of typical turtles (the latter being the original "tortoise shell"). During the Late Cretaceous, it appears that the trionychids occupied the range of aquatic habitats from freshwater to salt water, and their remains are quite common in the brackish water deposits of the estuaries that also were inhabited by *Deinosuchus*.

With this host of turtles, the southeastern coasts of the Late Cretaceous United States must have been similar in that respect to modern (but pre-European) North America. Sea turtles have been especially hard hit by the past 400 years of hunting, accidental by-catching during conventional net fishing, accidental collisions with boat propellers, and marine pollution (e.g., strangulation and choking from plastic debris). Even during the days of the 18th- and 19th-century sailing ships, sea turtles were easily caught and exploited as living food storage for long voyages; they were chained, upside-down, on ship's decks for months, and killed as needed for meat. Nevertheless, despite all the impacts and past exploitation, one may walk along the Georgia or north Florida coasts today and see the nests and bones of hundreds to thousands of marine turtles. In nesting grounds along the coasts of southeast Texas and Mexico, marine turtle eggs are still collected for human food by the tens of thousands, where they are not protected. Premodern marine turtle populations were probably enormous, and this undoubtedly was true back to the Late Cretaceous, although the species were different.

Large turtles, both marine and freshwater, are aggressive and even formidable animals. Sea turtles are themselves predators, feeding on fish and invertebrates. (The largest living turtle, the leatherback, may weigh over ~300 kg, yet feeds exclusively on jellyfish; nevertheless, their beaks can reportedly kill human-size animals). In the modern world, there are few marine animals that are equipped to prey on an adult turtle the size of a loggerhead (*Caretta caretta*), green (*Chelonia mydas*), or leatherback (*Dermochelys coriacea*). The killer whales and most sharks are certainly large enough to disable the biggest turtles, but their teeth are too fragile to break or crush the shells, and at best (or one might say worst), they can only bite off limbs—an inefficient mode of predation. Nevertheless, many sea turtles are injured or killed by sharks (Budker 1971), and this is usually done by the very largest species—great white, mako, and tiger sharks—where the teeth are fairly strong by virtue of sheer size. In the case of tiger sharks (*Galeocerdo* species), the teeth are also relatively thick and low-crowned, and these are the sharks most frequently observed to be marine turtle-eaters. In contrast, freshwater turtles have many natural predators, in part because they are usually smaller than marine species. But in warmer areas of North America, larger alligators feed abundantly on freshwater turtles, including the largest snappers and soft-shelled turtles that may have shell lengths of 0.6 m and weights to ~30 kg.

Unlike modern times, during the later Mesozoic Era and Early Tertiary Period, several categories of marine reptiles were sufficiently large and equipped with the types of teeth necessary to be chelonivores (i.e., turtle predators). These reptiles include *Deinosuchus* and several additional types of crocodylians, plus some specialized mosasaurs (marine lizards) in the genus *Globidens*. The key to recognizing this mode of feeding is their globidentine teeth set into unusually strong jaws of sufficient length to encompass a turtle within the gape (Fig. 8.4). As with all fossil evidence, there is some degree of assumption that must be made in claiming a predator–prey nexus. But the presence of rounded, crushing teeth in a large predator that co-occurs in the same habitats as hard-shelled turtles certainly suggests such a relationship. This is all the more true considering that most marine vertebrates (i.e., aside from turtles) tend to be thin skinned and lightly boned because marine water is buoyant, and marine vertebrate bodies do not require the strong skeletal support structures of land animals. There are exceptions, such as some heavily scaled fishes (e.g., trunkfish, sturgeon) and crocodylians themselves; nevertheless, since the later Mesozoic, the most abundant hard-boned marine vertebrates have been turtles. Therefore, when we find fossils of a marine animal adapted to feeding on hard-boned animals, the first assumption to make is that it was a turtle-eater. Certainly invertebrates, especially clams, may be hard shelled, and many ancient and modern predators are known to eat them. But before presuming a fossil predator ate invertebrates, we must consider the predator's body size relative to the prey because it seems improbable that a large, active predator would adapt to eating animals with magnitudinally smaller body sizes. (The exceptions are the gigantic plankton eaters, such as basking sharks and baleen whales, where the huge sizes of their oral cavities allow passively engulfing masses of water and microscopic prey —and *Deinosuchus* was obviously not a planktivore!)

In a previous discussion in Chapter 7, I briefly discussed a Tertiary dyrosaurid mesoeucrocodylian, *Phosphatosaurus*, which is presumed to have been chelonivorous on the basis of its robust jaws and blunt, reinforced teeth (Buffetaut 1979). *Phosphatosaurus* was also a very big animal that coexisted with many large marine turtles. The Late Cretaceous alligatoroids, *Albertochampsa* and *Brachychampsa*, from the western United States, are likewise considered to have been turtle-eaters (Carpenter and Lindsey 1980). This assumption is based largely on the same criteria as discussed above; however, these were smaller crocodylians and presumably were feeding on the typically smaller freshwater turtles (including soft-shelled species) with which they were associated in freshwater deposits in Montana. *Brachychampsa*, which is the better-known form, features a lower jaw reinforced with a lingual shelf, presumably to better handle the pressures of turtle crushing. In their discussion, Carpenter and Lindsey (1980) noted that one may distinguish comparably sized mollusk-crushing predators from chelonivores by the detailed nature of the teeth. Looking over a broad range of mollusk-crushing vertebrates, including such divergent forms as late Mesozoic sharks, Triassic reptiles called placodonts, and modern sea otters, one observes that mollusk eaters have very flat, closely spaced

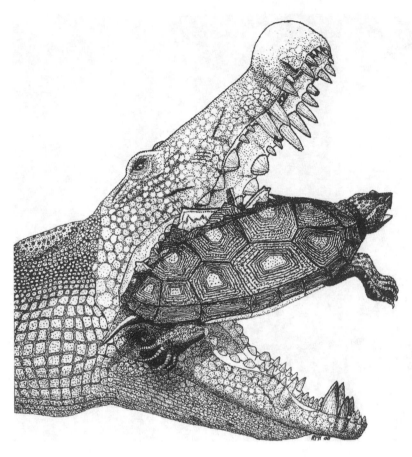

Figure 8.4. Deinosuchus *with a turtle in its jaws. This assumed feeding method employed the heavy rear teeth as the shell crushers because they are in the region of maximum jaw strength and leverage.*
Drawing by Ron Hirzel.

teeth with shallow roots. In contrast, turtle-eaters have rounded, separated teeth with deep roots, and by that criterion, the globidentine crocodylians were clearly chelonivores. Carpenter and Lindsey also observed that chelonivores are fairly rare through geological history, and today only some crocodylians (and humans) are clearly such.

Even though the idea of ancient crocodylians as chelonivores is hardly a novelty, the body elements indicating chelonivory are present in *Deinosuchus* to an even greater degree than in nearly all other crocodylians, including the heavy-toothed forms *Phosphatosaurus* and *Brachychampsa*. But in addition to tooth and body morphology, with *Deinosuchus*, we have even more compelling evidence of chelonivory. In my own collections, I have turtle fossils bearing tooth marks that are reasonably attributed to *Deinosuchus*, as well as some *Deinosuchus* teeth showing signs of having bitten some very hard surfaces. These specimens include many fragments of shells from the pleurodire turtle *Bothremys*, all from the Blufftown Formation at the western Georgia–eastern Alabama border region. These turtle shell fragments bear the tooth marks of feeding by an animal with large, blunt teeth (Fig. 8.5), and given the known animals in the associated fossil assemblage, that animal would have to be *Deinosuchus*.

Figure 8.5. Representative shell fragments from the side-necked marine turtle Bothremys *showing blunt bite marks. The depressions in the bone are approximately 1 cm wide. These and many similar fossils come from the Blufftown Formation at Hannahatchee Creek, Georgia.*

Among eight *Bothremys* specimens with bite marks, two are the most revealing of their origin, and they are especially interesting and contrasting because one shows that the turtle died from the event, and the other proves that the turtle survived. The fatal bite mark (Fig. 8.6) is a circular puncture through a fragment of dorsal shell in one of the neural (vertebral) bones. The puncture is about 1 cm across, and the shape is most indicative of its physical cause because the hole is smaller on the external surface and expands on the inner side. This inwardly expanding shape shows that extreme pressure caused this puncture (rather than, say, the boring of a snail or other invertebrate). The roundness of the hole shows that a round, globidentine tooth was the cause, and this in turn points to *Deinosuchus* in the fauna. One may wonder why there is no evidence of cracking around the puncture, but experiments I have done in my laboratory with fresh, wet, flat bone (e.g., pork ribs), show that hard pressure at a single point does not crack the surrounding bone but rather penetrates cleanly. Once the bone is dried (i.e., in a dead animal), it will crack much more easily. On the specimen shown in Figure 8.6, the expanded internal opening of the bite hole is the result of tooth pressure essentially blowing out the inner surface.

The second revealing *Bothremys* specimen is a water-worn bone fragment from a large turtle (with bone ~2 cm thick), probably a piece

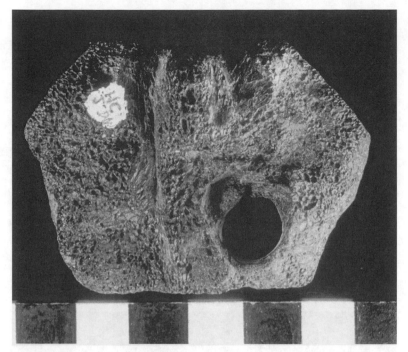

Figure 8.6. Bothremys *neural bone with a round perforation. (Neurals are the midline bones of the turtle carapace that are equivalent to dorsal vertebrae in typical vertebrates.) This view is the internal surface, and it is noteworthy that the perforation expands inward, indicating that external pressure caused the hole. Scale in centimeters. Specimen from the Blufftown Formation, Russell County, Alabama. Courtesy of G. Dent Williams.*

of the large first costal bone (where the anterior limbs attach). It has a large, round depression (Fig. 8.7), which penetrates about 1.0 cm thickness. When I collected this specimen in western Georgia, I noticed that the fossil bone enclosing the hole was unusual for *Bothremys,* but it took some close study to realize why. Rather than revealing the typically irregular and open internal bone fabric of turtle carapace bone, the bone lining the interior of the hole in this specimen is dense and tightly organized. The texture is reminiscent (but not nearly as compact) as the outer layer of the shell (i.e., the cortical bone). I believe this specimen shows that the hole is a bite that was healing before the turtle died! The regrown bone was an attempt by the turtle's body to regenerate cortical bone around the hole after a large *Deinosuchus* clamped down on the shell and somehow lost it. I assume the *Deinosuchus* was large (for the southeastern United States) because the hole is quite large in diameter, measuring ~2.0 cm in diameter. But this doesn't reveal the true dimensions of the tooth that caused it. I have inserted a midjaw tooth from a large *Deinosuchus* into the specimen (Fig. 8.7) to illustrate that the hole does indeed conform to the same outline. The hole is deep enough to actually contain and support the tooth, but the tooth is really much larger than the diameter of the hole left by the tip.

There is another line of evidence that shows *Deinosuchus* was a turtle-eater in the southeast. I have seen several hundred *Deinosuchus* teeth from various local fossil collections from the Late Cretaceous, and among those, about 30% of the posterior, low-crowned teeth show either broken or severely rounded tips. In this context, "rounded"

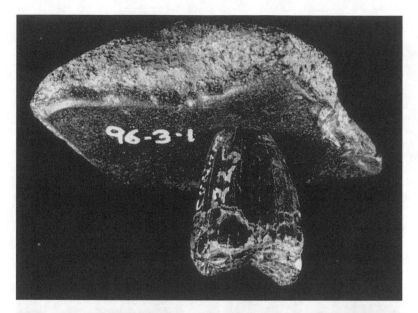

Figure 8.7. (below) Bothremys plastron fragment, with an apparently healed crocodylian bite.
(above) The same specimen showing that a worn posterior Deinosuchus fits quite well into the hole. Both specimens from the Blufftown Formation, Hannahatchee Creek, Georgia.

means from wear, not just the original rounded profile. It is fairly simple to distinguish the two because a worn tooth tip in *Deinosuchus* shows concentric circles of enamel and dentine, whereas the original external surface shows the rugosities, which give the species name, and a uniformly colored surface. I also have seen a smaller percentage (~15%) of the anterior, high-crowned *Deinosuchus* teeth with similar wear or breakage. The broken-tipped teeth, front and rear, are especially noteworthy because they are often broken off cleanly and sharply at an oblique angle (Fig. 8.8). This type of brittle breakage could have occurred in some specimens after fossilization, especially because the teeth are quite hard when mineralized in phosphatic salts (see pp. 36,

116). But the large number of similar breaks on so many teeth suggests otherwise: that they broke that way during life. Farlow and Brinkman (1994) discussed and illustrated some tyrannosaurid teeth from the Judith River Formation with their tips shattered in patterns similar to the damage I observe in many *Deinosuchus*. They tentatively concluded that the breakage resulted from biting the hard bones of prey species. This therefore means that these strong crocodylian teeth broke off on some hard or resistant substance. Among all the prey animals one may conceive, once again, thick-shelled turtles would seem to be the most obvious cause of the broken teeth.

As I discussed in the chapter opening, there seem to be differing feeding behaviors among the smaller eastern and larger western deinosuchids. The types of evidence for chelonivory that are found in the southeast are so far unknown in the west. This is not completely unexpected because isolated *Deinosuchus* teeth and turtle fossils are much more rare in the west, although better-associated skulls and skeletons are present, and we would be hard-pressed to prove or disprove the idea that chelonivory was going on. However, from multiple evidences, it is apparent that dinosaurs were a major prey of the western crocodylians.

Figure 8.8. Posterior and midjaw Deinosuchus *teeth, with sheared tips indicating breakage on hard-shelled prey. All specimens from the Blufftown Formation, Hannahatchee Creek, Georgia.*

The Dinosaur Connection

Finally, in this last section of the last chapter of the book, we come to the topic—let's be honest—that may have prompted you to pick it up in the first place. This is the compelling image of huge crocodylians killing and feeding on dinosaurs. In the first chapter, I presented a hypothetical scenario of *Deinosuchus* killing a small theropod in a southeastern Coastal Plain swamp. The story was based on both some direct evidence and a lot of inference, as will be discussed. Also to come: the same and even stronger direct evidence and inference on dinosaur-eating may be applied to the *Deinosuchus* populations of the western United States.

The concept of dinosaur predation has a long association with *Deinosuchus*, dating back to the earliest discoveries in Montana and especially in Big Bend, with the original description of the "*Phobosuchus riograndensis*" type material. No reasonable paleontologist, observing the huge dimensions of *Deinosuchus* bones and teeth, could fail to consider the likely association of these massive predators with the dinosaurs found in the same deposits. In recent years, we have accumulated some explicit evidence that these assumptions were correct, and further, we can show that even the smaller *Deinosuchus* found in the eastern side of the continent were feeding on dinosaurs. Because there is no recognized term for the concept, rather than coin a horrific one such as "dinophagy" or "dinivory," I will simply refer to the process from this point on as "dinosaur-eating."

Before examining the direct evidence for dinosaur-eating, we should consider the question of why *Deinosuchus* evolved into a lineage of crocodylians with the size and equipment to prey on dinosaurs. In all evaluations of paleontological evolutionary events, there is the "*post hoc* problem": that the events were ancient but the observations are modern. It is obvious that most *Deinosuchus* were large enough to kill and eat at least some dinosaurs, but we cannot know whether their large sizes were *achieved* because dinosaurs were abundant as potential prey, or whether already large sizes *allowed* them to prey on large dinosaurs. The first idea verges on Lamarckianism (i.e., the flawed concept of evolution by acquired traits), but even so, it is not completely unjustifiable. For example, the opportunity to feed on large prey could have been naturally selected (i.e., by Darwinian reasoning) for success by larger members of a preexisting *Deinosuchus* population. Thus, they would outbreed the smaller individuals and drive the population toward larger size. This scenario assumes that an earlier *Deinosuchus* population had other prey opportunities before they evolved sufficient size to prey on dinosaurs, and because many crocodylians are big animals, this assumption is quite reasonable. An alternative hypothesis is that *Deinosuchus*, with all or most of its characteristics, including large size, evolved in whole cloth as a dinosaur predator from an unknown alligatoroid ancestor. This hypothesis is also reasonable because in modern evolutionary studies, we recognize that relatively rapid evolutionary events are to be expected following the model of punctuated equilibrium formulated by Gould and Eldredge (1993). By this reasoning, we might consider that minor behavioral or physical changes (e.g., better ability to maneuver on land), or a change in access to prey (e.g., the coevolution of more aquatic and marine shore-based dinosaurs, such as hadrosaurs) could open up new opportunities for the earliest *Deinosuchus* population. In turn, these crocodylians would rapidly evolve into dinosaur predators, which would naturally tend to grow large because they were amply fed and because larger size enabled better killing of larger dinosaurs.

I believe that the first argument has better support, but both are reasonable. My bias toward the first alternative perhaps stems from my closer association with the eastern United States *Deinosuchus* fauna, and for the following reasons. As discussed in Chapter 6, the oldest eastern *Deinosuchus* fossils are slightly older than the first to appear in the west, and in fact, most of the eastern *Deinosuchus* are a few million years older than the western. This suggests (but does not prove) that the eastern population was probably the first and shows the more basal characters. Extending this reasoning, because the eastern *Deinosuchus* are the smaller forms, this is the presumed ancestral condition. So far, this argument holds together well, and it also supports the idea that early *Deinosuchus* populations preyed on a variety of animals, including, of course, turtles. Whether these early eastern *Deinosuchus* ate dinosaurs rarely or frequently is unclear, but we know they did so occasionally because we have direct evidence of this behavior.

We can go farther out on a hypothetical limb and speculate that it is the very set of adaptations to feeding on hard-boned prey that enabled *Deinosuchus* to evolve into a dinosaur killer in its younger, western population. Because one of the most distinctive sets of features of *Deinosuchus* is the heavy teeth and strong jaws, these same agents of extraordinary crushing power would be excellent exaptations (i.e., adaptations applied to a new purpose) for killing small- to medium-sized dinosaurs. I described rather vividly in Chapter 1 how a *Deinosuchus* might dispatch a medium-sized theropod dinosaur by crushing its lower back vertebrae and by tearing the carcass apart in the same manner that modern crocodiles do with large, hoofed mammals. Because many dinosaurs had heavily scaled, thick skin (which we know from the mummified remains of several hadrosaurs), the sharper, piercing teeth of typical crocodylians would not be the most effective killing tools. But if early *Deinosuchus* employed their millstone-heavy rear crushing teeth as the primary killing tools, then they could paralyze sizable dinosaurs with bites across the neck, back, and pelvis (Fig. 8.9). Obviously, they could also crush the lower legs of bipedal dinosaurs wading through marshes, which would bring upright animals into better reach of the low-slung crocodylians.

Going even further in speculation, I suggest that the occasional adventures into dinosaur-eating by eastern *Deinosuchus* evolved into a primary feeding behavior when the ancestral population migrated to the western side of the Interior Seaway. As *Deinosuchus* expanded its range to the western side, the founder population encountered a habitat with a large dinosaur population, which was probably more abundant

Figure 8.9. Reconstruction of a coastal setting in Late Cretaceous Big Bend as a Deinosuchus *attacks a pair of* Kritosaurus. *Drawing by Ron Hirzel.*

and diverse than that of the east (Schwimmer 1997b; but see Russell 1997 for a contrasting opinion). For example, ceratopsids (i.e., advanced horned dinosaurs) are unknown in the east, whereas they are very abundant in the west. And roughly speaking, for every dinosaur species known in the east, there are five dinosaur species, of a given geological age, in the west. Once a viable *Deinosuchus* breeding population was established on the western side of the Western Interior Seaway, I believe that the enhanced opportunities for feeding on abundant large dinosaurs selected for increasingly larger individuals of *Deinosuchus*. This feeding opportunity, and the advantages for such diets that large size would confer, rapidly led to the evolution of the huge *Deinosuchus* individuals typical of southwest Texas, Montana, and Wyoming.

To test this hypothesis, I would hope to see certain lines of fossil evidence. The first matter to document is whether there is material or other evidence that *Deinosuchus* began feeding on dinosaurs in the eastern United States. The second would be to find evidence of dinosaur-eating by the crocodylians in the west. As I will show, we have both sets of evidence available. The scenario presented in Chapter 1 was based on such hard evidence, although some details were extrapolations: I based it on a single theropod bone from New Jersey that appears to be crocodile-chewed, as well as on the report by Gallagher (1993) of carnivorous dinosaur teeth showing evidence of having been partly digested by crocodylian stomach acids. And I based it on observations made when I first began studying the ancient vertebrate population dynamics of the southeastern Coastal Plain (and before focusing on *Deinosuchus*), noting that really large carnivorous dinosaurs are absent, implying that some other animal occupied their predatory role.

Figure 8.10. Theropod limb fragment with numerous blunt bite marks, interpreted as Deinosuchus *feeding traces. Scale in centimeters. Specimen from the Ellisdale Site, Marshalltown Formation, Monmouth County, New Jersey. Courtesy of New Jersey State Museum.*

The first evidence is the most direct and compelling. This is a fragment of theropod limb bone, probably a tibia, in the collections of the New Jersey State Museum. The specimen came from the Ellisdale Site (Chapter 5), and what makes it so interesting, both by itself and in what it represents, is that it has been mutilated with crushing, blunt bites (Fig. 8.10). It is about 16 cm long, originally cylindrical, and about 6 cm in original diameter (but clearly deformed from crushing). It is obviously a theropod bone, on the basis of the dense cortex and smooth surfaces lining the medullary (central) cavity, which is an osteological characteristic linking theropods and birds. From the size of the fragment, assuming it is from the midshaft region of a larger theropod leg, the animal would be immature and weigh about 300–400 kg. There are comparable smaller theropod specimens known from North Carolina (Baird and Horner 1979) and in Georgia and Alabama (Schwimmer et al. 1993); however, at the present level of knowledge of the theropods of the eastern United States, it is not possible to assign this bone to genus. (There is a paper currently in peer review by T. D. Carr, and others, including me, identifying a medium-sized, Campanian-age tyrannosauroid theropod from Alabama as a new, primitive genus. I assume that many, if not all of the comparably sized theropod remains from the Campanian of the southeastern United States may be from this same animal. Whether this genus ranged as far north as New Jersey is undeterminable at present, but it is plausibly the source of the bone in question.)

The theropod fragment in Figure 8.10 has a great many blunt bite marks on all surfaces, including both deep depressions and shallow pits. Indeed, it is so thoroughly mangled that it resembles a dog's worn chew toy. The originally round cross section of the bone has been distorted into a nearly square configuration; this distortion apparently occurred before the bone was fossilized, almost certainly from the crushing action of whatever caused the bite marks. I believe these are clearly crocodylian bite marks because they are blunt and show no signs of serrations (as would appear in bites from other theropods). The nature of the bite marks documents tremendous crushing power, both because of the distortion of the bone and because of the pressure required to impress the teeth deeply into the surface. The presence of marks on all surfaces of the bone further shows that these do not result from boring mollusks or other organisms because they could not inhabit both sides of the bone at once as it lay on the sea bottom. Putting these ideas together, I deduce that this bone was the result of *Deinosuchus* killing or scavenging a small theropod. It also makes quite a bit of sense that we would, in particular, see a lower leg bone chewed up from such an event. In a discussion of eastern Cretaceous dinosaur occurrences (Schwimmer 1997b), I observed that the majority of dinosaur fossils we find regionally are either the ends of limbs or the ends of tails, for several reasons largely based on a "bloat-and-float" model of transportation of carcasses from the marine shore. Both shore-dwelling *Deinosuchus* and patrolling sharks apparently had roles in this pattern of dinosaur occurrences. The crushed theropod specimen is among the few that document the crocodylian role in dinosaur taphonomy (i.e., the larger term used for changes that occur between living organisms and preserved fossils).

Another and related line of evidence comes from the same locality in New Jersey. Gallagher (1993, 1995) discussed theropod teeth from the Ellisdale Site and attributed them to the odd theropod *Dryptosaurus aquilunguis*. Most interestingly, he briefly noted that some of these teeth had their exterior enamel partially dissolved away, which Gallagher attributed to the effects of either crocodylian or shark stomach acids. Essentially, he was claiming that these theropods (mostly juveniles) had either been eaten by a *Deinosuchus,* with the teeth excreted only partly digested, or scavenged by a shark after the dead animal floated out to sea. Gallagher's observations and ideas are quite reasonable, not only because of all the implications of dinosaur-eating we have considered so far, but also because there is prior knowledge and study on the effects of crocodylian stomach acid in bone digestion. In his study of *Crocodylus niloticus* in Africa, Cott (1961) examined the gut contents of recently killed crocodylians and evaluated the states of dissolution of vertebrate bones. Fisher (1981) specifically focused on the excretions of *Alligator mississippiensis* and made detailed observations about tooth and bone solution. Fisher observed that bones and teeth passing through the digestive system of alligators were characteristically demineralized, having been largely stripped of calcareous materials by the stomach acids. He also noted that the organic framework of such bones was largely preserved because these compounds are not equally soluble in hydrochloric acid. The effects of acid dissolution on teeth (versus bones) were especially pronounced because tooth enamel is a largely calcareous mineral; thus, teeth passing through a crocodylian are stripped of their external enamel, a fact readily observable even in fossil remains. On the basis of Fisher's observations, and putting all the information together, it seems that Gallagher (1993) was almost certainly discussing theropod teeth weathered out from *Deinosuchus* feces.

Rounding out the arguments that *Deinosuchus* began its dinosaur-eating behavior in the eastern United States are some general observations I have made on the bigger picture of Late Cretaceous animal populations of the east, especially in the southeast. We have many reports of theropod dinosaur remains on the Coastal Plains of the Atlantic and Gulf, usually consisting of single teeth or single bones. A few more complete remains are known ranging up to a two-thirds complete skeleton; but among all of these, none is from a very large animal. To my knowledge, the largest theropod known anywhere in the Late Cretaceous of the east would be less than 7.0 m long and would weigh no more than 1000 kg. This is certainly not an insignificant size for a carnivorous animal, but it is still approximately one-third smaller in all dimensions than comparable theropods of the same age known from the western side of the Interior Seaway. Most of the theropod remains in the east are from much smaller, immature individuals, and these observations lead me to speculate that some ecological effect was preventing larger theropods from thriving in the coastal habitats that formed the regional fossil record. Two additional observations tend to support this impression. First, in the same eastern deposits we find teeth and bones from full-sized hadrosaur dinosaurs. One has to wonder why large dinosaur herbivores are present (and even common) whereas large dinosaur carnivores are absent. Second, in the only known Mid-Cretaceous fossil deposit in the east, the Arundel Formation in Maryland (Lipka 1998) (also an apparently nearshore marine unit), I have observed some teeth from large theropods that exceed the size of any known from later in the Cretaceous in the same region. These observations suggest that a fierce competitor arose during the Late Cretaceous to challenge the theropods as dominant predators in the marine-shore areas. Their presence apparently precluded theropods from maintaining a successful breeding population of adults near the eastern coasts of the United States. Even more interesting, it appears that these dominant predators were able to kill and feed on theropods themselves—predator-on-predator behavior. I speculate that the theropod fossils we do find are young individuals that wandered down to the coasts from more favorable habitats farther inland; during the Late Cretaceous, these would be lowland Appalachian regions that left no fossil deposits. (Perhaps teenage dinosaurs were no more mindful of danger than are teenage humans?) Of all the known elements of Late Cretaceous coastal fossil assemblages, even before supporting evidence appeared, the giant crocodylian *Deinosuchus* was the logical candidate as the top predator in the regional coastal ecosystems. This phenomenon is especially evident in the eastern Gulf of Mexico region, where the largest population of *Deinosuchus* seems to have lived, and where we also find almost entirely small theropod remains among the known fossils. The correlation is probably not spurious.

There are many localities containing *Deinosuchus* fossils in the eastern United States (see Appendix C), but only a few are on the western side of the Seaway. Among the localities in the west containing their fossils, *Deinosuchus* remains are rare in Montana and Wyoming and only moderately common in the Big Bend region of southwest Texas. It is impossible to judge the size of the *Deinosuchus* population in most parts of the west, except perhaps in Big Bend, where they seem to have been fairly abundant. Until better information comes along, the parsimonious assumption we should make is that *Deinosuchus* was indeed rare in the more northern parts of the western interior coast and somewhat more common southward, on the Texas coast. One implication of these assumptions is that we may not expect to see a sizable amount of fossil evidence of *Deinosuchus* feeding comparable to that we observe in the east, simply because fewer (although generally larger individual) crocodylians may have been present. Fortunately, a single (but impressive) set of specimens provides some crocodylian feeding evidence we would hope to see, and as predicted, it appears on dinosaur remains. Although this is only a limited amount of information, it is the most clear indication that is known of the interactions of big, blunt-toothed crocodylians and dinosaurs. It also appears, for the first time, on the remains of large, herbivorous dinosaurs.

I heard reports of feeding traces on hadrosaur bones from several colleagues shortly after I began research on the paleobiology of *Deinosuchus*; and in early 1998, I was shown photographs that were the source of these reports. These photos showed a single hadrosaur caudal (tail) vertebra with many blunt bite marks that appeared to be of the type one might expect from a large crocodylian. The owner of this specimen was Ken Barnes, an active fossil collector and outdoors-expedition entrepreneur, located in the small (and most unusual) town of Terlingua, Texas, at the southwest edge of Big Bend National Park. I visited the park and vicinity in the summer of 1999 and was shown the specimens that were part of a fossil exhibit that Mr. Barnes maintains. He houses this exhibit in an old school bus, which was at the time parked at his home in the Terlingua Ghost Town. (This is a resurrected 1940s cinnabar mining encampment, which has been partly renovated by hardy individuals who choose to live in the most remote village in the lower 48 states of the United States.) The school bus was stuffed with interesting fossils, mostly from the Aguja and underlying Pen Formations of the Late Cretaceous; the sites from which Barnes extracted the fossils were (by law) outside the national park, and because of the commercial value of dinosaur fossils, localities such as Barnes' are traditionally kept confidential. Therefore, I cannot be absolutely certain that the following assumptions made from my brief observations (it was ~40°C in the bus) are correct as stated, but I have a lot of confidence that they are as reported here.

There were three hadrosaur caudal vertebrae with bite marks in the Barnes' collections, and they appeared to be from two different animals. I base this assumption both from the accounts by Mr. Barnes and on the proportional sizes and positions of these bones. The single caudal vertebra from what we will call the first site is an anterior, whereas the two from the second site are midcaudals. All came from the tails of average-sized hadrosaurs, and they probably can be referred to the most common regional genus *Kritosaurus*. The specimen I had seen previously photographed (Fig. 8.11) was the most obviously and aggressively bitten, with deep marks on all surfaces, especially the cranial and caudal sides (i.e., front and back) of the centrum. The bite marks are round in cross section, ranging up to about 0.9 cm across, and are pressed into the bone up to about 5 mm in depth. They range greatly in size and depth, and there are an impressive number of them—for example, I counted 19 distinct holes on the cranial (anterior) side, with many depressions that could be additional bites. In general, these holes

Figure 8.11. Hadrosaur vertebra from the Aguja Formation, Big Bend Texas, showing numerous blunt bite marks interpreted as Deinosuchus *feeding traces. (left) Anterior view. (right) Right lateral view. Similar marks are on the opposite two sides of the specimen. Specimen courtesy of Ken Barnes, Terlingua, Texas.*

Figure 8.12. Hadrosaur caudal vertebra from the Aguja Formation, Big Bend, Texas, showing two apparent bite marks (near center and upper right). This specimen is not from the same individual as Figure 8.11. Specimen courtesy of Ken Barnes.

resemble the single bite marks on the *Bothremys* fragments illustrated in Figure 8.5, but most are slightly larger. From all signs, and from the obvious inferences discussed here, I assume that these bite traces were left by the blunt rear teeth of a *Deinosuchus*. I noted that there was no evidence of cracking around the margins of the bites, in common with the *Bothremys* specimen, showing that the bone was fresh and wet at the time it was chewed. The other two vertebrae in Barnes' collection each has a single depression. Both depressions are larger (Fig. 8.12) than any on the first specimen described, and both are somewhat eroded. According to Mr. Barnes, these two bones were found in association and come from a different locality than the first specimen. The bite mark on one of these latter specimens is roughly similar to examples in the previous set in being round and fairly deep. The third specimen has a wide, shallow mark with an irregular outline, which is the only one of these traces that appears to have involved a significant amount of bone shattering or erosion after the animal's death.

I can make tentative assumptions about the events that produced these fossil remains, which are different for the two sets of hadrosaur tailbones. The anterior caudal vertebra from the first site was obviously worked over for a considerable amount of time after being separated from the rest of the dinosaur, which is amply proven by the abundance of bite marks on all four sides. This makes particularly good sense, considering that in the anterior region of a duck-billed dinosaur the tail was large and fleshy, and there would be a great mass of muscle and other tissue (e.g., ossified tendons) to be chewed. Indeed, were the specimen available for detailed study, I assume we could discover that there are pairs of upper and lower tooth holes impressed on opposite sides of the bone. It is not possible to prove whether or not this hadrosaur was dead before being attacked by the crocodylian, or if it was killed in the same time frame as the bite marks were impressed. The evidence that the bone was fresh while it was bitten (documented by the absence of cracks) somewhat supports the idea that it was a fresh kill; but bone can remain supple for a long time after an animal's death if covered by other tissue.

The two vertebrae from the second site seem to reflect a slightly different scenario. The presence of one bite apiece suggests they were part of a larger mass that was being chewed by the crocodylian, and that this unit had the flesh pulled off the bones (or was dropped) before the vertebrae received further bites. The fossils are too badly ablated (i.e., worn and missing edges) to judge whether the bitten surfaces represent opposite sides of the two vertebrae—that is, whether they are a front and back, as one would expect if they were in sequence, while the hadrosaur tail was being chewed up. Because these latter bite marks are less obvious than on the first specimen and are of such limited extent, I am not equally sure they are definitively the result of *Deinosuchus* activity.

To my knowledge, these bite-marked hadrosaur vertebrae are the only specimens that clearly or likely document crocodylian feeding on dinosaurs in the western United States. I spent part of a summer in Big Bend searching for more specimens and observed only one additional specimen showing possible evidence of *Deinosuchus* feeding activity. This was yet another hadrosaur tail vertebra, a large midcaudal, with a single, vague, round depression on the ventral margin. I consider this specimen significantly less compelling than even the poorer one in Barnes' collections. And beyond even this vague evidence, I was unable to find further proof of *Deinosuchus* dinosaur-feeding in Big Bend, nor have I heard further reports of such specimens. Nevertheless, I am not surprised that these are extraordinarily rare occurrences in the western Late Cretaceous fossil record; this follows from simple deduction, beginning with the assumption that the crocodylians were probably fairly rare.

One might imagine that with all the dinosaur fossils known in the western United States, and the obviously enormous number of predatory events that must have occurred during the dinosaur age, evidences of feeding would be abundant. But this is not the case, and there are only a handful of reports from all North American dinosaur sites showing feeding traces (e.g., Carpenter 1988; Erickson and Olson 1996; Erickson et al. 1996; Fiorello 1991). If we now consider that activity by

Figure 8.13. The top predators of the southeastern coast refuse to share their shore dinner. Drawing by Ron Hirzel.

Deinosuchus would be restricted to the nearshore areas, we must further reduce the likely opportunities to find such evidence by the proportional reduction of area that might reveal the bones. If we finally reduce that remaining amount by the proportion of dinosaur sites that overlap with the nearshore regions known to be occupied by *Deinosuchus,* the expectations of finding such evidence becomes very small indeed. In fact, it is pretty much limited to the Aguja Formation in Big Bend, and we can better appreciate the significance of the few specimens known from there.

From the few bite traces we have in Big Bend, it would be purely speculative to judge the size and other details of the crocodylians that left the feeding signs. The holes show that the teeth were blunt, and to penetrate dinosaur vertebrae as observed, they must have been strong. Because the bite marks were left by only the tips of teeth, it is guesswork to specify the dimensions of the teeth involved, but we can reasonably speculate that only a very large crocodylian would feed on a structure as massive as a hadrosaur tail (Fig. 8.13). Among the crocodylians in the Aguja Formation in Big Bend (Lehman 1997), including *Leidyosuchus* and *Goniopholis,* only *Deinosuchus* is a plausible source of the limited dinosaur-feeding evidence. Although one could wish to see more abundant direct evidence of dinosaur predation by *Deinosuchus,* I believe the case is adequately made in Big Bend, as it is in the eastern United States.

Appendix A. Geologic Time and Subdivisions of the Late Cretaceous

Throughout the book, there are frequent references to subdivisions of the Cretaceous geological time period, including "Late Cretaceous," "Campanian Age," and others. For easy reference, and to place these in perspective, the following chart includes the detailed subdivisions of the Late Cretaceous Epoch, with appropriate dates for the boundaries. The reader who wishes to put these time units into the larger perspective of all geological time can refer to the charts in any textbook of historical geology. Absolute age data (i.e., years) come from Harland et al. (1990); Obradovich (1993); and Bralower et al. (1995).

EPOCH	AGE		DATES (MYR BEFORE PRESENT)
	Maastrichtian	Late	66.5–65.4
		Early	74.0–66.5
	Campanian	Late	76.0–74.0
		Early	83.0–76.0
Late Cretaceous	Santonian	Late	85.0–83.0
		Early	86.5–85.0
	Coniacian		88.5–86.5
	Turonian	Late	89.5–88.5
		Early	90.5–89.5
	Cenomanian		97.0–90.5

Appendix B.
Glossary of
Anatomic and
Cladistic Terms

The following terms are defined in the narrow focus referring to croc-odylomorphs. They have been used more than once in the book and may not have been defined on the second or subsequent usages. I have not included formal taxonomic names, which may be found in Table 7.1 (p. 138) and in the index. The names of specific crocodylian bones are included here if they have been used in the text more than once; Figure 3.1 indicates the positions and shape of nearly all external skull bones in *Deinosuchus*.

In most cases, I have broken down compound usages to their root terms and it will be necessary to refer to both parts for a full definition: for example, for "temporal fenestrae," refer to "temporal" for the anatomical position and "fenestra (-ae)" for the structure. However, some usages are significant to this subject only in their compound form (e.g., "ghost lineages"), and I have defined them as such. Where a term has been used in both the singular and irregular plural forms, I have listed the singular with the plural suffix in parentheses. Similarly, some words have been used in both the noun and adjectival forms; in these cases, the noun is listed and the adjective is in parentheses.

acetabulum	The socket in the pelvis into which the femur fits.
alveoli	Sockets in the jaw bones into which the teeth fit.
amphicoelous	Condition in which the centrum of each vertebra is concave on both ends.
antorbital	In front of the eye socket.
apomorphy	A derived (newly evolved) character state.

articular	The bone of the lower jaw that articulates with the quadrate bone of the upper jaw.
astragalus	The tarsal bone that articulates with the tibia.
autapomorphy	A derived (newly evolved) feature unique to a single taxon.
basal	At the base of a lineage.
brevirostrine	Having a broad, relatively short snout.
calcaneum	The tarsal bone that articulates with the fibula.
caudal	Of the tail region.
centrum	The cylindrical main body of the vertebra.
cervical	Of the neck region.
chelonivore (-ivory)	Turtle-eating.
choanae	Internal nostrils located in the roof of the mouth.
clade	A lineage within a phylogenetic tree, usually consisting of an ancestor and all of its descendants.
cladogram	A branching diagram showing hypothetical relationships of taxa.
cloaca	The common aperture into which the digestive, urinary, and reproductive systems open.
cololite	Fossilized intestinal casts.
coprolite	Fossilized excrement.
cortex (cortical)	The outer layers of a structure (c.g., a bone).
costal	Relating to the ribs.
crown group	In cladistics, the highest taxa including extant species, descended from a common ancestor.
crurotarsal	Type of ankle joint in which the line of flexure runs diagonally between the astragalus and the calcaneum, allowing both hinging and rotational movement.
crus	The lower portion of the hind leg between the knee and the ankle.
dentary	The tooth-bearing bone of the mandible.
derived	An evolutionarily advanced feature.
'derm. See *osteoderm.*	
dorsal	The upper side, or nearest to the back.

ectopterygoid	A pair of bones contributing to the lateral portion of the palate.
epicontinental sea	A shallow sea covering the interior continental crust—for example, the Western Interior Seaway of the Late Cretaceous.
eustachian tube	A canal which connects the middle ear with the pharynx.
exaptation. See *preadaptation.*	
fenstra (ae)	A larger opening in bones (literally, "window").
ghost lineage	A hypothetical lineage for which there is no fossil record.
globidentine	Teeth with low, very rounded crowns, often with thick enamel.
ilium	The dorsal blade of the pelvis, which articulates with the sacral vertebrae.
ischium	Ventral, backward projecting part of the pelvis.
homoplasy	The appearance in different lineages of similar structures not inherited from a common ancestor.
kinetic (kinesis)	Bone connections capable of movement.
labial	The outside surface of the jawbones.
lateral	Away from the midline of the body, or to the side.
lingual	Closest to the tongue; the inside surface of the jawbones.
longirostrine	Having a long, narrow snout.
mandible	The lower jaw, composed of two sides (right and left).
maxilla (maxillary)	The upper jawbone.
monophyletic	A clade that contains an ancestor and all of its descendants.
nasal	The dorsal pair of bones that extend down the middle of the skull and make up much of the length of the rostrum.
occlusion	The fit between teeth of the upper and lower jaws.
orbit	The eye socket.
osteoderm	A bony plate embedded in the skin.
otic	Of the ear.

paraphyletic	A group which contains an ancestor and some, but not all, of its descendants.
parsimony	Use of the simplest assumption in the interpretation of data (also called "Occam's Razor").
pericontinental sea	A marine incursion over the coastal plain on the exterior of the continent, as on the Atlantic and Gulf coasts of the Late Cretaceous.
phylogeny	The evolutionary relationships among groups of organisms.
plantigrade	Type of stance in which the entire foot contacts the ground.
plesiomorphy	A primitive character state inherited from an ancestor.
polyphyletic	Groups which are descended from different ancestral lineages; not a natural group.
preadaptation	A mutation which may originally be of no selective advantage, but which may become selectively advantageous in a new or changing environment.
premaxilla	The paired bones that form the anteriormost part of the skull and upper jaw.
process	A flange or extension of a bone.
procoelous	Condition in which the centrum of each vertebra is concave anteriorly.
pterygoid	The largest and most medially situated of the ventral skull bones, immediately posterior to the palatines.
pubis	The anterior, ventral part of the pelvis.
quadrate	One of the posterior, ventral bones of the skull, which, with the articular bone of the mandible, forms the jaw joint.
rostrum	An elongation of the skull forward of the orbits, containing the nasals, maxilla, and premaxilla.
scute	A bony dermal plate (see *osteoderm*).
secondary palate	A ventral shelf of bone, formed below the skull roofing bones, in the rostrum.
sister group	Taxa which stem from the same phylogenetic node but are not colinear with a given clade.
skull table	The high, flattened, posterodorsal region of the skull roof in Crocodyliformes.

splenials	Bones which form part of the ventral and inner surface of the mandible.
stem group	Taxa which may be placed between the point where an ancestral group splits into two sister groups, to the point at which a further split gives rise to crown groups.
surangular	Forms the upper posterior part of the outer surface of the mandible in reptiles.
suture	An immobile joint between two bones.
symbiosis (-es)	Species interactions, commonly assumed to be mutually beneficial.
symplesiomorphy	A "primitive" character state shared by two or more taxa.
synapomorphy	A derived character state shared by two or more taxa.
taphonomy	The study of the processes involved in fossilization.
tarsus	Bones of the ankle region.
temporal	Of the posterior, dorsal skull region.
tetrapod	A vertebrate having four limbs (i.e., amphibians and higher).
ventral	The lower side, or closest to the abdomen.
ziphodont	Having serrated, carnivorous-dinosaur–like teeth.
zygapophysis	A process on the vertebra which takes part in linking the vertebrae to each other.

Appendix C.
List of
Deinosuchus
Localities

The sites here are listed alphabetically by states, and within the states, they are arranged by the abundance of known specimens. Many more sites containing *Deinosuchus* fossils undoubtedly exist, but the following are known to me at the time of this writing.

Alabama

(Eastern) Stream valleys in Russell County, with many collection sites occurring along the banks. The streams include High Log, Hatchechubbee, and North and Middle Forks of Cowikee Creeks. All are in the Blufftown Formation. Fossils include isolated vertebrae, teeth, osteoderms, and several partial mandibles.

Barbour County, stream banks along the South Fork of Cowikee Creek, in the upper Blufftown Formation. Fossils are isolated teeth and osteoderms.

(Central) Lowndes County, on the west bank of the Alabama River at the Army Corps of Engineers Lock and Dam. A single specimen came from the Mooreville Formation consisting of a partial skull and complete right mandible (Fig. 5.6).

(Western) Greene County, near the town of West Greene, from the Mooreville Formation. The single *Deinosuchus* fossil assemblage is the remains of a juvenile specimen, lacking a skull but containing many bones and most of the osteoderms.

Georgia

(Western) Stewart County, Hannahatchee Creek, upper Blufftown Formation: isolated teeth, vertebrae, osteoderms, mandible and skull fragments, and *Deinosuchus*-bitten turtle bones.

Chattahoochee County, east bank of Chattahoochee River, lower Blufftown Formation: complete right mandible and ilium (Figs. 4.7, 7.13).

Mississippi

(Northeastern) Lee County, Tulip Creek near Tupelo, Coffee Sand Formation: partial skull and partial left mandible (Fig. 5.7).

Montana

(Central) Fergus County, near Willow Creek, Judith River Formation: numerous osteoderms, vertebrae, ribs, and pubis (holotype of *Deinosuchus hatcheri*; Fig. 2.9).

New Jersey

(Central) Monmouth County, Ellisdale Township, Marshalltown Formation: isolated teeth, vertebrae, osteoderms, and *Deinosuchus*-bitten theropod bone (Fig. 8.10).

North Carolina

(Southeastern) Bladen County, Phoebus Landing and vicinity, Black Creek Formation: isolated teeth, vertebrae, mandible fragments, osteoderms, and vertebrae (holotype of "*Polyptychodon*" *rugosus*).

Texas

(Southwestern) Brewster County, numerous sites in Big Bend National Park and vicinity, Aguja Formation: several skulls, mandibles, vertebrae, isolated teeth, osteoderms, postcranial bones (holotype of *Phobosuchus riograndensis*; Figs. 2.1, 2.4–2.6, 2.8, 6.5).

Wyoming

(Northcentral) Washakie County, Mesaverde Formation, single cervical or dorsal osteoderm in the collection of the University of Wyoming. I observed the specimen, courtesy of William Wahl (Casper Museum), and it is undoubtedly *Deinosuchus*.

Unpublished reports of *Deinosuchus* remains from central-western Wyoming are common knowledge. Isolated osteoderms in the U.S. National Museum were attributed to *Deinosuchus hatcheri* by R. W. Howell of the U.S. Geological Survey and came from the Parkman Formation, in the Salt Creek Oil Field at coordinates T 41°N, R 79°W.

References

Alcober, O. 2000. Redescription of the skull of *Saurosuchus galilei* (Archosauria, Rauisuchidae). *Journal of Vertebrate Paleontology* 20: 302–316.

Armstrong-Hall, J. G. 1999. Amiid and ptyctodontid [*sic*] remains in the Cretaceous fauna of Missouri. *Journal of Vertebrate Paleontology* 19(3, abstracts): 30A.

Baird D., and P. M. Galton. 1981. Pterosaur bones from the Upper Cretaceous of Delaware. *Journal of Vertebrate Paleontology* 1(1): 67–71.

Baird, D., and J. R. Horner. 1979. Cretaceous dinosaurs of North Carolina. *Brimleyana* 2: 1–28.

Beaumont, C., G. M. Quinlan, and G. S. Stockmal. 1993. The evolution of the Western Interior Basin: Causes, consequences and unsolved problems. In W. G. E. Caldwell and E. G. Kauffman (eds.), *Evolution of the Western Interior Basin,* pp. 97–118. Special Paper 39. St. John's, Newfoundland: Geological Association of Canada.

Benton, M. J., and J. M. Clark. 1988. Archosaur phylogeny and the relationships of the Crocodylia. In M. J. Benton (ed.), *The Phylogeny and Classification of the Tetrapods,* Vol. 1, pp. 295–338. London: Clarendon Press.

Bird, R. T. 1985. *Bones for Barnum Brown.* Fort Worth, Tex.: Texas Christian University Press.

Bralower, T. J., R. M. Leekie, W. V. Sliter, and H. R. Thierston. 1995. An integrated Cretaceous microfossil biostratigraphy. In *Geochronology Time Scales and Global Stratigraphic Correlation,* pp. 65–79. Tulsa, Okla.: SEPM Special Publication 54.

Brinkman, D. R. 1990. Paleoecology of the Judith River Formation (Campanian) of Dinosaur Provincial Park, Alberta, Canada: Evidence from vertebrate microfossil localities. *Palaeogeography, Palaeoclimatology, Palaeoecology* 78: 37–54.

Bryan, J. R., D. L. Frederick, D. R. Schwimmer, and W. Seisser. 1991. First dinosaur record from Tennessee: A Campanian hadrosaur. *Journal of Paleontology* 65: 696–697.

Brochu, C. A. 1997a. A review of "*Leidyosuchus*" (Crocodyliformes, Eusuchia) from the Cretaceous through Eocene of North America. *Journal of Vertebrate Paleontology* 17: 679–697.

———. 1997b. Synonymy, redundancy, and the name of the crocodile stem-group. *Journal of Vertebrate Paleontology* 17: 448–449.

————. 1999. Phylogeny of the Alligatoroidea. *Journal of Vertebrate Paleontology* 19(Suppl. 2): 9–102.

————. 2000. Phylogenetic relationships and divergence timing of *Crocodylus* based on morphology and the fossil record. *Copeia* 2000: 657–673.

Buckley, G. A., C. A. Brochu, D. W. Krause, and D. Pol. 2000. A pug-nosed crocodyliform from the Late Cretaceous of Madagascar. *Nature* 405: 941–944.

Budker, P. 1971. *The Life of Sharks.* New York: Columbia University Press.

Buffetaut, E. 1979. The evolution of the Crocodilians. *Scientific American* 241: 130–144.

————. 1982. Radiation évolutive, paléoécologie et biogéography des crocodiliens mésosuchians. *Mémoires de la Société Géologique de France,* n.s., 60(142): 1–88.

Buffetaut, E., and B. Taquet. 1977. The giant crocodilian *Sarcosuchus* in the Early Cretaceous of Brazil and Niger. *Palaeontology,* 20: 203–208.

Busbey, A. B., III. 1997. The structural consequences of skull flattening in crocodilians. In J. J. Thomason (ed.), *Functional Morphology in Vertebrate Paleontology,* pp. 173–192. Cambridge: Cambridge University Press.

Busbey, A. B., III, and T. M. Lehman (eds.), 1989. *Vertebrate Paleontology, Biostratigraphy and Depositional Environments, Latest Cretaceous and Tertiary, Big Bend Area, Texas.* Guidebook, Field Trips 1A–C. Society of Vertebrate Paleontology, 49th Meeting, Austin, Texas.

Byerly, G. R. 1991. Igneous activity. In A. Salvador (ed.), *The Geology of North America,* Vol. J, *The Gulf of Mexico Basin,* pp. 91–109. Boulder, Colo.: Geological Society of America.

Campbell, K. E., and C. D. Frailey. 1991. Summary of results of paleontological survey and collecting activities in Southwest Amazonia. *Journal of Vertebrate Paleontology* 11(3, abstracts): 19A–20A.

Cappetta, H. 1987. Chondrichthyes II: Mesozoic and Cenozoic Elasmobranchii. In H.-P. Schultze (ed.), *Handbook of Paleoichthyology,* Vol. 3B, p. 139. Stuttgart, Germany: Gustav Fischer Verlag.

Carpenter, K. 1988. Evidence of predatory behavior by *Tyrannosaurus.* In J. R. Horner (ed.), *International Symposium on Vertebrate Behavior as Derived from the Fossil Record,* n.p. Bozeman, MT: Museum of the Rockies, Montana State University.

Carpenter, K., and D. Lindsey. 1980. The dentary of *Brachychampsa montana* Gilmore (Alligatorinae, Crocodylidae), a Late Cretaceous turtle-eating alligator. *Journal of Paleontology* 54: 1213–1217.

Carroll, R. L. 1988. *Vertebrate Paleontology and Evolution.* New York: W. H. Freeman.

Case, G. R. 1991. Selachians (sharks) from the Tupelo Tongue of the Coffee Sand (Campanian, Upper Cretaceous) in Northern Lee County, Mississippi. *Mississippi Geology* 11: 1–8.

Case, G. R., and D. R. Schwimmer. 1988. Late Cretaceous fish from the Blufftown Formation (Campanian) in western Georgia. *Journal of Paleontology* 62: 290–301.

Chin, K. 1997. Coprolites. In P. J. Currie and K. Padian (eds.), *Encyclopedia of Dinosaurs,* pp. 147–150. New York: Academic Press.

Chin, K., and B. D. Gill. 1996. Dinosaurs, dung beetles, and conifers: Participants in a Cretaceous food web. *Palaios* 11: 280–285.

Chin, K., T. T. Tokaryk, G. M. Erickson, L. C. Calk. 1998. A king-size theropod coprolite. *Nature* 393: 680–682.

Chinnery, B. K., T. B. Lipka, J. I. Kirkland, J. M. Parrish, and M. K. Brett-Surman. 1998. Neoceratopsian teeth from the Lower to Middle Cretaceous of North America. In S. G. Lucas, J. I. Kirkland, and J. W. Estep (eds.), *Lower and Middle Cretaceous Terrestrial Ecosystems,* pp. 297–302. Albuquerque: New Mexico Museum of Natural History and Science, Bulletin 14.

Clark, J. M. 1996. Patterns of evolution in Mesozoic Crocodyliformes. In N. C. Fraser and H.-D. Sues (eds.), *In the Shadow of the Dinosaurs: Early Mesozoic Vertebrates,* pp. 84–97. Cambridge: Cambridge University Press.

Colbert, E. H. 1997. North American dinosaur hunters. In J. O. Farlow and M. K. Brett-Surman (eds), *The Complete Dinosaur,* pp. 24–33. Bloomington: Indiana University Press.

Colbert, E. C., and R. T. Bird. 1954. A gigantic crocodile from the Upper Cretaceous beds of Texas. *American Museum of Natural History Novitates* 1688.

Cope, E. D. 1869. (Remarks on . . . *Polydectes biturgidus.*) *Proceedings of the Academy of Natural Sciences, Philadelphia,* p. 192.

———. 1871. Observations on the distribution of certain extinct vertebrata in North Carolina. *Proceedings of the American Philosophical Society* 12: 210–216.

Coria, R. A., and L. Salgado. 1995. A new giant carnivorous dinosaur from the Cretaceous of Patagonia. *Nature* 377: 224–226.

Cott, H. B. 1961. Scientific results of an inquiry into the ecology and economic status of the Nile crocodile (*Crocodylus niloticus*) in Uganda and Northern Rhodesia. *Transactions of the Zoological Society of London* 29: 211–357.

Coulson, T. D., R. A. Coulson, and T. Hernandez. 1973. Some observations on the growth of captive alligators. *Zoologica* (summer): 47–52.

Cox, R. T. 1997. Possible track of the Bermuda Hotspot across the southeastern U.S. In *Abstracts with Programs of the Geological Society of America, Southeastern Meeting,* Vol. 29, p. 10. Boulder, Colo.: Geological Society of America.

Craig, L. E., A. G. Smith, and R. L. Armstrong. 1989. Calibration of the geologic time scale: Cenozoic and Late Cretaceous glauconite and nonglauconite dates compared. *Geology* 17: 830–832.

Cruikshank, A. R. I. 1979. The ankle joint in some early archosaurs. *South African Journal of Science* 75: 168–178.

Cumbaa, S. L., and T. T. Tokaryk. 1993. Early birds, crocodile tears, and fish tales: Cenomanian and Turonian marine vertebrates from Saskatchewan. *Journal of Vertebrate Paleontology* 13(3, abstracts): 31A–32B.

Daly, E. 1992. *A List, Bibliography, and Index of the Fossil Vertebrates of Mississippi.* Jackson: Mississippi Department of Environmental Quality, Office of Geology, Bulletin 128.

Davies, K. L., and T. M. Lehman. 1989. The WPA quarries. In A. B. Busbey III and T. M. Lehman (eds.), pp. 32–42. *Vertebrate Paleontology, Biostratigraphy and Depositional Environments, Latest Cretaceous and Tertiary, Big Bend Area, Texas.* Guidebook, Field Trips 1A–C. Society of Vertebrate Paleontology, 49th Meeting, Austin, Texas.

Dockery, D. T., III. 1997. Late Cretaceous volcanic activity and associated uplift at Jackson, Mississippi. In *Abstracts with Programs of the Geological Society of America, Southeastern Meeting,* Vol. 29(3), p. 14. Boulder, Colo.: Geological Society of America.

Dockery, D. T., III, and J. C. Marble. 1998. Seismic Stratigraphy of the Jackson Dome. *Mississippi Geology* 19(3): 29–43.

Dodson, P. 1996. *The Horned Dinosaurs.* Princeton, N.J.: Princeton University Press.

Dowling, H. G., and P. Brazaitis. 1966. Size and growth in captive crocodilians. *International Zoo Yearbook* 6: 265–270.

Druckenmiller, P. S., A. J. Daun, J. L. Skulan, and J. C. Pladziewicz. 1993. Stomach contents in the Upper Cretaceous shark *Squalicorax falcatus. Journal of Vertebrate Paleontology* 13(3, abstracts): 33A–34A.

Dyke, G. J. 1998. Does archosaur phylogeny hinge on the ankle joint? *Journal of Vertebrate Paleontology* 18: 558–562.

Echols, J. 1972. Biostratigraphy and reptile faunas of the upper Austin and Taylor Groups (Upper Cretaceous) of Texas. Ph.D. dissertation. Norman, Okla.: University of Oklahoma.

Emmons, E. 1858. *Report of the North Carolina Geological Survey.* Raleigh, N.C., pp. 219–221, figs. 38–39.

Erickson, B. R. 1972. *Albertochampsa langstoni,* gen. et sp. nov., a new alligator from the Cretaceous of Alberta. *Scientific Publications, Science Museum of Minneapolis,* n.s., 2: 1–13.

———. 1976. *Osteology of the Early Eusuchian Crocodile* Leidyosuchus formidabilis, *sp. nov.* Monograph 2, *Paleontology.* St. Paul: Science Museum of Minnesota.

Erickson, G. M., and C. A. Brochu. 1999. How the "terror crocodile" grew so big. *Nature* 398: 205–206.

Erickson, G. M., and K. H. Olson. 1996. Bite marks attributable to *Tyrannosaurus rex*: Preliminary description and implications. *Journal of Vertebrate Paleontology* 16: 175–178.

Erickson, G. M., S. D. Van Kirk, J. Su, M. E. Levenston, W. E. Caler, and D. R. Carter. 1996. Bite-force estimation for *Tyrannosaurus rex* from tooth-marked bones. *Nature* 382: 706–708.

Estes, R. 1964. Fossil vertebrates from the Late Cretaceous Lance Formation eastern Wyoming. *University of California Publications in Geological Sciences (Berkeley)* 49.

Farlow, J. O., and D. L. Brinkman. 1994. Wear surfaces on the teeth of tyrannosaurs. In G. D. Rosenberg and D. L. Wolberg (eds.), *Dino Fest,* pp. 164–175. Paleontological Society Special Publications 7.

Fiorello, A. 1989. The vertebrate fauna from the Judith River Formation (Late Cretaceous) of Wheatland and Golden Valley Counties, Montana. *Mosasaur* 4: 127–148.

———. 1991. Prey bone utilization by predatory dinosaurs. *Palaeography, Palaeoclimatology, Palaeoecology* 88: 157–166.

Fisher, D. M. 1981. Crocodilian scatology, microvertebrate concentrations, and enamel-less teeth. *Paleobiology* 7: 262–275.

Fleagle, J. 1988. *Primate Evolutionary Anatomy.* New York: Academic Press.

Frazier, W. J., and Schwimmer, D. R. 1987. *Regional Stratigraphy of North America.* New York: Plenum Press.

Frey, E. 1984. Aspects of the biomechanics of crocodilian terrestrial locomotion. In W.-E. Reif and F. Westphal (eds.), *Third Symposium on Mesozoic Terrestrial Ecosystems,* pp. 93–97. Tübingen: Short Papers.

Gallagher, W. B. 1984. Paleoecology of the Delaware Valley region, part II: Cretaceous to Quaternary. *Mosasaur* 2: 9–43.

Gallagher, W. G. 1993. Cretaceous–Tertiary mass extinction event in the North Atlantic Coastal Plain. *Mosasaur* 5: 75–154.

———. 1995. Evidence for juvenile dinosaurs and dinosaurian growth stages in the Late Cretaceous deposits of the Atlantic Coastal Plain. *Bulletin of the New Jersey Academy of Science* 40: 5–8.

Gauthier, J. A. 1984. A cladistic analysis of the higher systematic categories of the Diapsida. Ph.D. dissertation. University of California, Berkeley.

Gauthier, J. A., and K. Padian. 1985. Phylogenetic, functional and aerodynamic analyses of the origin of birds and their flight. In M. K. Hecht et al. (eds.), *Beginnings of Birds*. pp. 185–197. Eichstätt, Germany: Freunde des Jura-Museums.

Gervais, P. 1876. Crocodile gigantesque fossile au Brésil. *Journal de Zoologie* 5: 233–236.

Gomani, E. 1997. A Crocodyliforme from the Early Cretaceous dinosaur beds, northern Malawi. *Journal of Vertebrate Paleontology* 17: 280–294.

Gottfried, M. D. 1997. Reconstructing and mounting a skeleton of the giant "megatooth" shark *Carcharodon megalodon*. *Journal of Vertebrate Paleontology* 17(3, abstracts): 50A.

Gould, S. J. 1992. *Bully for Brontosaurus*. New York: W. W. Norton.

Gould, S. J., and N. Eldredge. 1993. Punctuated equilibrium comes of age. *Nature* 366: 223–227.

Gow, C. E. 2000. The skull of *Protosuchus haughtoni,* an early Jurassic Crocodylifiorme from southern Africa. *Journal of Vertebrate Paleontology* 20: 49–56.

Greer, A. E. 1974. On the maximum total length of the salt-water crocodile (*Crocodylus porosus*). *Journal of Herpetology* 8: 381–384.

Haq, B. U., J. Hardenbol, and P. R. Vail. 1987. Chronology of fluctuating sea levels since the Triassic. *Science* 235: 1156–1166.

Harland, W. B., et al. 1990. *A Geologic Time Scale*. Cambridge: Cambridge University Press.

Harris, W. B. 1982. Rubidium–strontium glaucony ages, southeastern Atlantic Coastal Plain, USA. In G. S. Odin (ed.), *Numerical Dating in Stratigraphy*, pp. 591–602. New York: Wiley.

Hass, C. A., M. A. Hoffman, L. D. Densmore III, and L. R. Maxson. 1992. Crocodilian evolution: Insights from immunological data. *Molecular Phylogenetics and Evolution* 1: 193–201.

Hay, O. P. 1902. *Bibliography and Catalogue of the Fossil Vertebrata of North America*. Washington, D.C.: U.S. Geological Survey Bulletin 179.

Hay, W. W., D. L. Eicher, and R. Diner. 1993. Physical oceanography and water masses in the Cretaceous Western Interior Seaway. In W. G. E. Caldwell and E. G. Kauffman (eds.), *Evolution of the Western Interior Basin*, pp. 297–318. Special Paper 39. St. John's, Newfoundland: Geological Association of Canada.

Heckert, A. B., and S. G. Lucas. 1999. A new aetosaur (Reptilia: Archosauria) from the Upper Triassic of Texas and the phylogeny of aetosaurs. *Journal of Vertebrate Paleontology* 19: 50–68.

Hicks, J. F., J. D. Obradovich, and L. Tauxe. 1995. A new calibration point for the Late Cretaceous time scale: The $^{40}Ar/^{39}Ar$ isotopic age of the C33r/C33n geomagnetic reversal for the Judith River Formation (Upper Cretaceous), Elk Basin, Wyoming, USA. *Journal of Geology* 103: 243–256.

Holland, W. J. 1909. *Deinosuchus hatcheri,* a new genus and species of crocodile from the Judith River beds of Montana. *Annals of the Carnegie Museum* 6: 281–294.

Horner, J. R. 1989. The Mesozoic terrestrial ecosystems of Montana. In *Montana Geological Society Field Conference Guidebook*, pp. 153–162.

Holtz, T. R., Jr. 1994. The phylogenetic position of the Tyrannosauridae:

Implications for theropod systematics. *Journal of Paleontology* 68: 1100–1117.

Hua, S., and E. Buffetaut. 1997. Crocodylia, Part IV. In J. Callaway and E. L. Nicholls (eds.), *Ancient Marine Reptiles,* pp. 357–374. New York: Academic Press.

Hungerbühler, A. 2000. Heterodonty in the European phytosaur *Nicrosaurus kapffi* and its implications for the taxonomic utility and functional morphology of phytosaur dentitions. *Journal of Vertebrate Paleontology* 20: 31–48.

Iordansky, N. N. 1973. The skull of the Crocodylia. In C. Gans and T. S. Parsons (eds.), *Biology of the Reptilia,* pp. 201–262. London: Academic Press.

Kase, T., P. A. Johnston, A. Seilacher, J. B. Boyce. 1998. Alleged mosasaur bite marks on Late Cretaceous ammonites are limpet (patellogastropod) scars. *Geology* 26: 947–950.

Kauffman, E. G., and W. G. E. Caldwell. 1993. The Western Interior Seaway in space and time. In W. G. E. Caldwell and E. G. Kauffman (eds.), *Evolution of the Western Interior Basin,* pp. 1–30. Special Paper 39. St. John's, Newfoundland: Geological Association of Canada.

Kranz, P. M. 1989. *Dinosaurs in Maryland.* Baltimore: Maryland Geological Survey Educational Series 6.

Lamb, J. P., Jr. 1997. A Late Cretaceous land bridge across the Gulf of Mexico Basin. *Journal of Vertebrate Paleontology* 17(3, abstracts): 59A.

———. 1998. *Lophorhothon,* an iguanodontian, not a hadrosaur. *Journal of Vertebrate Paleontology* 18(3, abstracts): 59A.

Langston, W., Jr. 1960. The dinosaurs. Part 6 of The vertebrate fauna of the Selma Formation of Alabama. *Fieldiana, Geology Memoirs* 3(6): 313–361.

———. 1966. *Mourasuchus* Price, *Nettosuchus* Langston, and the family Nettosuchidae (Reptilia: Crocodilia). *Copeia* 1966: 882–885.

———. 1974. Nonmammalian Comanchean tetrapods. *Geoscience and Man* 8: 77–102.

Langston, W., Jr., B. Standhardt, and M. Stevens. 1989. Fossil vertebrate collecting in the Big Bend. In A. B. Busbey III and T. M. Lehman (eds.), *Vertebrate Paleontology, Biostratigraphy and Depositional Environments, Latest Cretaceous and Tertiary, Big Bend Area, Texas,* pp. 11–21. Guidebook, Field Trips 1A–C. Society of Vertebrate Paleontology, 49th Meeting, Austin, Texas.

Lauginiger, E. M. 1984. An Upper Campanian vertebrate fauna from the Chesapeake and Delaware Canal, Delaware. *Mosasaur* 2: 141–149.

———. 1988. *Cretaceous Fossils from the Chesapeake and Delaware Canal: A Guide for Students and Collectors.* Newark: Delaware Geological Survey Special Publication 18.

Lawson, D. A. 1975. Pterosaur from the latest Cretaceous of West Texas: Discovery of the largest flying creature. *Science* 187: 947–948.

Lee, Y.-N. 1997. The Archosauria from the Woodbine Formation (Cenomanian) in Texas. *Journal of Paleontology,* 71: 1147–1156.

Lehman, T. M. 1987. Late Maastrichtian paleoenvironments and dinosaur biogeography in the Western Interior of North America. *Palaeogeography, Palaeoclimatology, Paleaeoecology* 60: 189–217.

———. 1989. Overview of Late Cretaceous sedimentation in Trans-Pecos Texas. In A. B. Busbey III and T. M. Lehman (eds.), *Vertebrate Paleontology, Biostratigraphy and Depositional Environments, Latest Cre-*

taceous and Tertiary, Big Bend Area, Texas, pp. 23–25. Guidebook, Field Trips 1A–C. Society of Vertebrate Paleontology, 49th Meeting, Austin, Texas.

———. 1997. Late Campanian dinosaur biogeography in the Western Interior of North America. In D. Wolberg, E. Stump, and G. D. Rosenberg (eds.), *Dinofest International Proceedings,* pp. 223–240. Philadelphia: Academy of Natural Sciences.

Leidy, J. 1870. Remarks on ichthyorudiolites and on certain fossil Mammalia. *Proceedings of the Academy of Natural Sciences, Philadelphia* 21: 12–13.

Lipka, T. R. 1998. The affinities of the enigmatic theropods of the Arundel Clay Facies (Aptian), Potomac Formation, Atlantic Coastal Plain, Maryland. In S. G. Lucas, J. I. Kirkland, and J. W. Estep (eds.). Lower and Middle Cretaceous Ecosystems. *New Mexico Museum of Natural History and Science Bulletin* 14: 229–234.

Manning, E. M., and D. T. Dockery. 1992. *A Guide to the Frankstown Vertebrate Fossil Locality (Upper Cretaceous), Prentiss County, Mississippi.* Jackson: Mississippi Department of Environmental Quality, Office of Geology Circular 4.

Markwick, P. J. 1998a. Fossil crocodilians as indicators of Late Cretaceous and Cenozoic climates: Implications for using palaeontological data in reconstructing palaeoclimate. *Palaeogeography, Palaeoclimatology, Palaeoecology* 137: 205–271.

———. 1998b. Crocodilian diversity in space and time: The role of climate in paleoecology and its implication for understanding K/T extinctions. *Paleobiology* 24: 470–497.

McIntosh, J. S., M. K. Brett-Surman, and J. O. Farlow. 1997. Sauropods. In J. O. Farlow and M. K. Brett-Surnam (eds.), *The Complete Dinosaur,* pp. 264–290. Bloomington: Indiana University Press.

McNulty, C. L., and B. H. Slaughter. 1969. A vertebrate local fauna from the uppermost Woodbine Formation (Cenomanian) of Tarrant County, Texas. *Geological Society of America Special Paper 121,* pp. 404–405. Boulder, Colo.: Geological Society of America.

Mehl, M. G. 1941. *Dakotasuchus kingi,* a crocodile from the Dakota of Kansas. *Denison University Bulletin, Journal of the Science Labs* 36: 47–65.

Meyer, E. R. 1984. Crocodilians as living fossils. In N. Eldredge and S. M. Stanley (eds.), *Living Fossils,* pp. 105–131. New York: Springer-Verlag.

Meyer, R. L. 1974. Late Cretaceous elasmobranches from the Mississippi and East Texas Embayments of the Gulf Coastal Plain. Ph.D. dissertation. Dallas, Tex.: Southern Methodist University.

Miller, H. W. 1967. Cretaceous vertebrates from Phoebus Landing, North Carolina. *Proceedings of the Academy of Natural Sciences, Philadelphia* 119: 219–235.

———. 1968. Additions to the Upper Cretaceous vertebrate fauna of Phoebus Landing, North Carolina. *Journal of the Elisha Mitchell Scientific Society* 84: 467–471.

Molnar, R. E. 1988. Biogeography and phylogeny of the Crocodylia. In R. E. Jones et al. (eds.), *Fauna of Australia.* Canberra: Australia Biological Resources, Study 2A, pp. 344–348.

Molnar, R. E., and H. T. Clifford. 2000. Gut contents of a small ankylosaur. *Journal of Vertebrate Paleontology* 20: 194–196.

Mook, C. C. 1925. A revision of the Mesozoic Crocodilia of North America. *Bulletin of the American Museum of Natural History* 51: 319–432.

Nolf, D., and D. T. Dockery. 1990. Fish otoliths from the Coffee Sand (Campanian) of northeastern Mississippi. *Mississippi Geology* 10(3): 1–14.

Nopcsa, F. B. 1924. Über die Namen einiger brasillianischer fossiler Krokodile. *Centralblatt für Mineralogie, Geologie und Palaontologie* 12: 378.

Norell, M. A. 1989. The higher level relationships of the extant Crocodylia. *Journal of Herpetology* 23(4): 325–335.

Norell, M. A., J. M. Clark, and J. H. Hutchinson. 1994. The Late Cretaceous alligatoroid *Brachychampsa montana* (Crocodylia): New material and putative relationships. *American Museum of Natural History Novitates* 3116.

Obradovich, J. 1993. A Cretaceous time scale. In W. G. E. Caldwell and E. G. Kauffman (eds.), *Evolution of the Western Interior Basin,* pp. 379–396. Special Paper 39. St. John's, Newfoundland: Geological Association of Canada.

Odin, G. S. (ed.). 1982. *Numerical Dating in Stratigraphy, Part I.* New York: Wiley-Interscience.

Ortega, F., Z. Gasparini, A. D. Buscalioni, and J. O. Calvo. 2000. A new species of *Araripesuchus* (Crocodylomorpha, Mesoeucrocodylia) from the Lower Cretaceous of Patagonia. *Journal of Vertebrate Paleontology* 20: 57–76.

Parris, D. C. 1986. *Biostratigraphy of the Fossil Crocodile* Hyposaurus *Owen from New Jersey.* Trenton: New Jersey State Museum Investigation, Vol. 4.

Parris, D. C., B. S. Grandstaff, R. K. Denton Jr., W. B. Gallagher, C. De Temple, S. S. Albright, E. E. Spamer, and D. Baird. 1987. Taphonomy of the Ellisdale dinosaur site, Cretaceous of New Jersey. Final report on grant 3299-86. Washington, D.C.: National Geographic Society.

Parris, D. C., B. S. Grandstaff, B. L. Stinchcomb, and R. K. Denton Jr. 1988. Chronister: The Missouri dinosaur site. *Journal of Vertebrate Paleontology* 8(3, abstracts): 23A.

Parris, D. C., B. S. Grandstaff, R. K. Denton Jr., and J. L. Dobie. 1997. *Diplocynodon* (Alligatorinae) in the Cretaceous of eastern North America. *Journal of Vertebrate Paleontology* 17(3, abstracts): 69A.

Parrish, J. M. 1986. Locomotor adaptations of the hindlimb and pelvis of the Thecodontia. *Hunteria* 1(2): 1–35.

———. 1997. Evolution of the Archosauria. In J. O. Farlow and M. K. Brett-Surman (eds), *The Complete Dinosaur,* pp. 191–203. Bloomington: Indiana University Press.

Paul, G. 1988. *Predatory Dinosaurs of the World.* New York: Simon and Schuster.

———. 1997. Dinosaur models: The good, the bad, and using them to estimate the mass of dinosaurs, In D. L. Wolberg, E. Stump, and G. D. Rosenberg (eds.), *Dinofest International Proceedings,* pp. 129–154. Philadelphia: Academy of Natural Sciences.

Poag, C. W. 1997. The Chesapeake Bay bolide impact: A convulsive event in Atlantic Coastal Plain evolution. *Sedimentary Geology* 108: 45–90.

Ricqlès, A. de, J. R. Horner, and K. Padian. 1998. Growth dynamics of the hadrosaurid dinosaur *Maiasaura peeblesorum. Journal of Vertebrate Paleontology* 18(3, abstracts): 72A.

Ride, W. D. L., et al. 1985. *International Code of Zoological Nomenclature.* 3rd edition. Berkeley: University of California Press.

Romer, A. S. 1956. *Osteology of the Reptilia*. Chicago: University of Chicago Press.

———. 1966. *Vertebrate Paleontology*. Chicago: University of Chicago Press.

Ross, C. A. (ed.). 1989. *Crocodiles and Alligators*. New York: Facts on File.

Rowe, T., R. L. Cifelli, T. M. Lehman, and A. Weil. 1992. The Campanian Terlingua local fauna, with a summary of other vertebrates from the Aguja Formation, Trans-Pecos Texas. *Journal of Vertebrate Paleontology* 12(4), pp. 472–493.

Russell, D. A. 1988. *A Checklist of North American Marine Cretaceous Vertebrates Including Fresh Water Fishes*. Drumheller, Alberta: Occasional Papers of the Tyrrell Museum of Palaeontology 4.

———. 1993. Vertebrates in the Cretaceous Western Interior Sea. In W. G. E. Caldwell, and E. G. Kauffman (eds.) *Evolution of the Western Interior Basin,* pp. 665–680. Special Paper 39. St. John's, Newfoundland: Geological Association of Canada.

———. 1997. Notes on dinosaurs down South. In D. L. Wolberg, E. Stump, and G. D. Rosenberg (eds.), *Dinofest International Proceedings*, pp. 241–244, Philadelphia: Academy of Natural Sciences.

Schwimmer, D. R. 1986. Late Cretaceous fossils from the Blufftown Formation (Campanian) in Georgia. *Mosasaur* 3: 109–123.

———. 1995. East–west Late Cretaceous marine vertebrate provincialism: Evidence of biogeography or artifact of parasynchrony? *Abstracts with Programs, Geological Society of America*, 27(6, abstracts): A387.

———. 1997a. Predatory dominance of giant crocodiles on the Late Cretaceous Southeastern Coastal Plain. *Abstracts with Programs of the Geological Society of America, Southeastern Meeting,* Vol. 29(3), p. 68. Boulder, Colo.: Geological Society of America.

———. 1997b. Late Cretaceous dinosaurs in eastern USA: A taphonomic and biogeographic model of occurrences. In D. Wolberg, E. Stump, and G. D. Rosenberg (eds.), *Dinofest International Proceedings*, pp. 203–211. Philadelphia: Academy of Natural Sciences.

———. 1997c. Disparity of North American Late Cretaceous marine vertebrate faunas: Perhaps more artifactual than real. *Journal of Vertebrate Paleontology* 16(3, abstracts): 64A.

———. 1999. On the size of the giant crocodylian *Deinosuchus*. *Journal of Vertebrate Paleontology* 19(3, abstracts): 74A.

Schwimmer, D. R., and G. D. Williams. 1993. A giant crocodile from Alabama and observations on the paleobiology of southeastern Crocodilians. *Journal of Vertebrate Paleontology* 13(3, abstracts): 56A.

———. 1996. New specimens of *Deinosuchus rugosus,* and further evidence of chelonivory by Late Cretaceous eusuchian crocodiles. *Journal of Vertebrate Paleontology* 16(3, abstracts): 64A.

Schwimmer, D. R., K. Padian, and A. B. Woodhead. 1985. First pterosaur records from Georgia: Open marine facies, Eutaw Formation (Santonian). *Journal of Paleontology* 39(3): 674–679.

Schwimmer, D. R., G. D. Williams, J. L. Dobie, and W. G. Seisser. 1993. Late Cretaceous dinosaurs from the Blufftown Formation in western Georgia and eastern Alabama. *Journal of Paleontology* 67: 288–296.

Schwimmer, D. R., J. D. Stewart, and G. D. Williams. 1994. Giant fossil coelacanths of the Late Cretaceous in the eastern United States. *Geology* 2: 503–506.

Schwimmer, D. R., J. D. Stewart, and G. D. Williams. 1997a. *Xiphactinus vetus* and the distribution of *Xiphactinus* species in the eastern United States. *Journal of Vertebrate Paleontology* 17: 610–615.

Schwimmer, D. R., J. D. Stewart, and G. D. Williams. 1997b. Scavenging by sharks of the genus *Squalicorax* in the Late Cretaceous of North America. *Palaios* 12: 71–83.

Scotese, C. R., L. M. Gahagan, and R. L. Larson. 1988. Plate reconstructions of the Cretaceous and Cenozoic ocean basins. *Tectonophysics* 155: 27–48.

Scott, R. W. 1970. Paleoecology and paleontology of the lower Cretaceous Kiowa Formation, Kansas. *University of Kansas Paleontological Contributions* 52: 1–94.

Seilacher, A., C. Marshall, H. C. W. Skinner, and T. Tsuihiji. 2001. A fresh look at sideritic "coprolites." *Paleobiology* 27: pp. 7–13.

Sereno, P. 1991. Basal archosaurs: Phylogenetic relationships and functional implications. *Journal of Vertebrate Paleontology,* Memoir 2.

Sereno, P., and A. B. Arucci. 1990. The monophyly of crurotarsal archosaurs and the origin of bird and crocodile ankle joints. *Neues Jahrbuch für Geologie und Paläontologie,* Abhandlung 180: 21–52.

Sereno, P., D. B. Duthiel, M. Iarochene, H. C. E. Larsson, G. Lyon, P. M. Magwene, C. A. Sidor, D. J. Varrichio, and J. A. Wilson. 1996. Predatory dinosaurs from the Sahara and Late Cretaceous faunal differentiation. *Science* 272: 986–991.

Sereno, P. C., H. C. E. Larsson, C. A. Sidor, and B. Gado. 2001. The giant crocodyliform *Sarcosuchus* from the Cretaceous of Africa. *Science* 294: 1516–1519.

Sissingh, W. 1977. Biostratigraphy of Cretaceous calcareous nannoplankton. *Geologie en Mijnbouw* 57: 37–65.

Sohl, N. F., E. Martinez, R. P. Salméron-Ureña, and F. Soto-Jamarillo. 1991. Upper Cretaceous. In *The Geology of North America,* Vol. 3, *The Gulf of Mexico Basin,* pp. 205–244. Boulder, Colo.: Geological Society of America.

Steel, R. 1973. *Encyclopedia of Paleoherpetology,* Part 16, *Crocodylia.* Stuttgart: Fischer Verlag.

Taplin, L. E., and G. C. Grigg. 1981. Salt glands in the tongue of the estuarine crocodile *Crocodylus porosus. Science* 212: 1045–1047.

———. 1989. Historical zoogeography of the eusuchian crocodilians: A physiological perspective. *American Zoologist* 29(3): 885–902.

Thurmond, J. T. 1969. Notes on mosasaurs from Texas. *Texas Journal of Science* 21: 69–80.

Thurmond, J. T., and D. E. Jones. 1981. *Fossil vertebrates of Alabama.* Tuscaloosa: University of Alabama Press.

Vaughn, P. P. 1956. Second specimen of the Cretaceous crocodile *Dakotasuchus* from Kansas. *Transactions of the Kansas Academy of Science* 59: 379–381.

Webb, G., and C. Manolis. 1989. *Crocodiles of Australia.* Frenchs Forest, Australia: Reed Publisher.

Weishampel, D. B., and L. Young. 1996. *Dinosaurs of the East Coast.* Baltimore, Md.: Johns Hopkins University Press.

Welton, B. J., and R. F. Farish. 1993. *The Collector's Guide to Fossil Sharks and Rays from the Cretaceous of Texas.* Lewisville, Tex.: Before Time.

Whetstone, K. N. 1978. *Belonostomus,* sp. (Teleostei, Aspidorhynchidae) from the Upper Cretaceous Tombigbee Sand of Alabama, Part 4. *University of Kansas Paleontological Contributions* 89: 17–19.

Williamson, T. E. 1996. ?*Brachychampsa sealeyi,* sp. nov. (Crocodylia, Alligatoroidea), from the Upper Cretaceous (lower Campanian) Menefee Formation, northwestern New Mexico. *Journal of Vertebrate Paleontology* 16: 421–431.

Witmer, L. M. 1997. The evolution of the antorbital cavity of archosaurs: A study in soft-tissue reconstruction in the fossil record with an analysis of the function of pneumaticity. *Journal of Vertebrate Paleontology,* Memoir 3.

Woodward, A. R., J. H. White, and S. B. Linda. 1995. Maximum size of the alligator (*Alligator mississippiensis*). *Journal of Herpetology* 29: 507–513.

Wu, X.-C., and S. Chatterjee. 1993. *Dibothrosuchus elaphros,* a crocodylomorph from the Jurassic of China and the phylogeny of the Sphenosuchia. *Journal of Vertebrate Paleontology* 13: 58–89.

Wu, X.-C., D. B. Brinkman, and A. P. Russell. 1996. A new alligator from the Upper Cretaceous of Canada and the relationships of early eusuchians. *Palaeontology* 39: 351–375.

Zangerl, R. 1953. The vertebrate fauna of the Selma Formation in Alabama, part III: The turtles of the family Toxochelyidae; part IV: An advanced cheloniid sea turtle. *Fieldiana Geology Memoirs* 3b: 135–312.

DAVID R. SCHWIMMER,
professor of paleontology at Columbus State University in Georgia, is
an expert on the Late Cretaceous paleontology of the southeastern
United States. Author of many papers on Cretaceous vertebrates, he is
coauthor (with W. J. Frazier) of *Regional Stratigraphy of North America,* which won the award for Best Reference Book of the Year from the
Geoscience Information Society.